Preventive Maintenance

Preventive Maintenance

Joseph D. Patton, Jr.

 Instrument Society of America

PREVENTIVE MAINTENANCE

© Instrument Society of America 1983

Printed in the United States of America

Instrument Society of America
67 Alexander Drive
P.O. Box 12277
Research Triangle Park, NC 27709

Library of Congress Cataloging in Publication Data

Patton, Joseph D.
  Preventive maintenance.

  Includes index.
  1. Maintenance.   I. Title.
TS192.P34   1982   658.2'02   82-48557
ISBN 0-87664-718-2
ISBN 0-87664-639-9 (pbk.)

Design and book production
by Publishers Creative Services Inc., New York

Contents

List of Figures

List of Tables

Preface and Acknowledgments

Preventive maintenance means all actions intended to keep durable equipment in good operating condition and to avoid failures. A good preventive maintenance (PM) program is the heart of effective maintenance. Success is often a matter of degree. The proper balance that achieves minimal downtime and costs can be tenuous between preventive maintenance and corrective maintenance. Everything is going to fail at some time. PM can prevent those failures from happening at a bad time, can sense when a failure is about to occur and fix it before it causes damage, and can often preserve capital investments by keeping equipment operating as well as it did on the day it was installed.

Inept PM, however, can also cause problems. Whenever any equipment is touched, it is exposed to potential damage. It is excessively costly to replace components prematurely. Customers may perceive the PM activity as, "that machine is broken again." A PM program requires an initial investment of time, parts, people, and money. Payoff comes months later. While there is little question that a good PM program will have a high return on investment, many people are reluctant to pay now if the return is not immediate. PM supports a commitment to long-term life-cycle cost/total cost of ownership.

Emotions play a prominent role in preventive maintenance. We all realize that perceptions often receive more attention than facts. A good data system, either manual or computerized, is necessary to provide the facts that must guide PM. PM is a dynamic process. It must support variations in equipment, wear, environment, use, personnel, schedules, and material. Changes are taking place both in technology and in management. These changes both require and support preventive maintenance. Technology provides the tools, and management provides the direction for their use. Both are necessary for success. These concepts are equally applicable to equipment and facility maintenance and field service in commerce, government, and industry.

Acknowledgments

Thanks to Amby T. Upfold and the personnel of Polysar Limited, and to Joseph Zdun, National Service Manager, and his staff at Leeds & Northrup for their review and constructive suggestions. Participants in workshops "How to Design and Implement a Preventive Maintenance Program" may recognize their enhancements. PCI consultants who have had considerable positive impact include Lawrence S. Beale, Herbert O. Feldmann, Michael A. Felluca, and Mary Ann Bianchi. Beverly C. Phillips typed the manuscripts from electronic dictation and drafted many of the illustrations.

CHAPTER 1

Major Types of Maintenance

There are three main types of maintenance and three major divisions of preventive maintenance, as illustrated in Figure 1-1.

IMPROVEMENT MAINTENANCE

Picture these divisions as the five fingers on your hand. Improvement maintenance efforts to reduce or eliminate

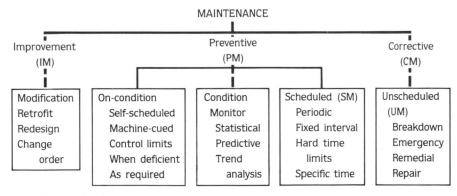

MAINTENANCE

Improvement (IM)	Preventive (PM)			Corrective (CM)
Modification	On-condition	Condition	Scheduled (SM)	Unscheduled
Retrofit	Self-scheduled	Monitor	Periodic	(UM)
Redesign	Machine-cued	Statistical	Fixed interval	Breakdown
Change	Control limits	Predictive	Hard time	Emergency
order	When deficient	Trend	limits	Remedial
	As required	analysis	Specific time	Repair

Figure 1-1
Structure of Maintenance

entirely the need for maintenance are like the thumb, the first and most valuable digit. We are often so involved in maintaining that we forget to plan ahead and eliminate the need at its source. Reliability engineering efforts should emphasize elimination of failures that require mainte- nance. This is an opportunity to preact instead of react.

For example, many equipment failures occur at in- board bearings that are located in dark, dirty, inaccessible locations. The oiler does not lubricate those bearings as often as he lubricates those that are easy to reach. That is a natural tendency. One can consider reducing the need for lubrication by using permanently lubricated, long-life bearings. If that is not practical, at least an automatic oiler could be installed. A major selling point of new auto- mobiles is the elimination of ignition points that require replacement and adjustment, introduction of self-adjust- ing brake shoes and clutches, and extension of oil-change intervals.

CORRECTIVE MAINTENANCE

The little finger in our analogy to a human hand repre- sents corrective (emergency, repair, remedial, un- scheduled). At present, most maintenance is corrective. Repairs will always be needed. Better improvement main- tenance and preventive maintenance, however, can reduce the need for emergency corrections. A shaft that is ob- viously broken into pieces is relatively easy to maintain because little human decision is involved. Troubleshooting and diagnostic fault detection and isolation are major time consumers in maintenance. When the problem is obvious, it can usually be corrected easily. Intermittent failures and hidden defects are more time-consuming but with diagnostics the causes can be isolated and then corrected. From a preventive maintenance perspective, the problems and causes that result in failures provide the targets for elimination by PM. The challenge is to detect insipient problems before they lead to total failures and to correct the defects at the lowest possible cost. That leads us to the middle three fingers—the branches of preventive maintenance.

PREVENTIVE MAINTENANCE

On-Condition

On-Condition maintenance is done when equipment needs it. Inspection through human senses or instrumentation is necessary, with thresholds established to indicate when potential problems start. Human decisions are required to establish those standards in advance so that inspection or automatic detection can determine when the threshold limit has been exceeded. Obviously, a relatively slow deterioration before failure is detectable by condition monitoring, whereas rapid, catastrophic modes of failure may not be detected. Great advances in electronics and sensor technology are being made.

Also needed is a change in human thought process. Inspection and monitoring should disassemble equipment only when a problem is detected. The following are general rules for on-condition maintenance:

—Inspect critical components.
—Regard safety as paramount.
—Repair defects.
—If it works, don't fix it.

Condition Monitor

Statistics and probability theory are the basis for condition monitor maintenance. Trend detection through data analysis often rewards the analyst with insight into the causes of failure and preventive actions that will help avoid future failures. For example, stadium lights burn out within a narrow range of time. If 10 percent of the lights have burned out, it may be accurately assumed that the rest will fail soon and should, most effectively, be replaced as a group rather than individually.

Scheduled

Scheduled, fixed interval PM should generally be used only if there is opportunity for reducing failures that cannot be detected in advance, or if dictated by production requirements. The distinction should be drawn between fixed interval maintenance and fixed interval inspection that may detect a threshold condition and initiate condition monitor PM. Examples of fixed interval PM include 7,500-mile oil changes and 12,000-mile spark plug changes on a car, whether it needs the changes or not. This may be very wasteful since all equipment and their operating environments are not alike. What is right for one situation may not be right for another.

SUMMARY

The way we think about PM holds the opportunity for great improvement. Preventive maintenance can provide major benefits if it is properly applied and if it truly prevents failures, reduces costs and downtime, and improves uptime, productivity, and profits.

The five-finger approach to maintenance emphasizes elimination and reduction of maintenance need wherever possible, inspection and detection of pending failures before they happen, repair of defects, monitoring of performance conditions and failure causes, and accessing the equipment on a fixed interval basis only if no better means exist.

CHAPTER 2

Advantages and Disadvantages

On the balance, preventive maintenance (PM) has many advantages. It is beneficial, however, to overview the advantages and disadvantages so that the good may be improved and the negative reduced. Note that in most cases the advantages and disadvantages vary with the type of PM used. Use of on-condition or condition monitor techniques is usually better than fixed intervals.

ADVANTAGES

Management Control

Unlike repair maintenance, which must react to failures, PM can be planned ahead. This means "preactive" instead of "reactive" management. Work loads may be scheduled so that equipment is available for PM at reasonable times.

Overtime

Overtime can be reduced or eliminated. Surprises are reduced. Work can be performed when convenient.

Work Load

Work loads can be balanced to either spread the demand over the available resources, or to hire additional personnel and equipment to meet the demand.

Equipment Uptime

While PM may require an investment of as many maintenance hours as were required previously for corrective maintenance, equipment should certainly perform better and with much higher availability when it is needed. It is a truism that failures are rarely found until equipment is put to use. PM, done properly, will often detect failures that have occurred but that would not otherwise be found until that equipment is needed for a rush job.

Production

Naturally, production will be happy because downtime, shutdowns, scheduling, and personnel problems will be reduced. Access to equipment is often restricted to specific times dictated by production requirements. PM helps assure best possible use of revenue producing functions.

Standardization

The "one best way" to do PM tasks should be determined. Because of the repetitious nature of PM, the procedures may be improved upon and skills may be finely honed. Maximum learning should be established early if the same persons are consistently used to conduct PM. Like any task done frequently with proper guidance, PM can reach a high level of productivity. This also permits more accurate planning because the times will evolve into a relatively narrow target range. Costs, like most insurance policies, are predetermined within small limits.

Parts Inventories

Since PM permits planning of which parts are going to be required and when, those material requirements may be anticipated to be sure they are on hand for the event. A smaller stock of parts is required in organizations that emphasize PM compared to the stocks necessary to cover breakdowns that would occur when PM is not emphasized.

Standby Equipment

With high demand for production and low equipment availability, reserve, standby equipment is often required in case of breakdowns. Some backup may still be required with PM, but the need and investment will certainly be reduced.

Safety and Pollution

If there are no PM inspections or built-in detection devices, equipment can deteriorate to a point where it is unsafe or may spew forth pollutants. Performance will generally follow a saw-tooth pattern, as shown in Figure 2-1, which does well after maintenance and then degrades until the failure is noticed and it is brought back up to a high level. A good detection system catches degrading performance before it ever reaches too low a level.

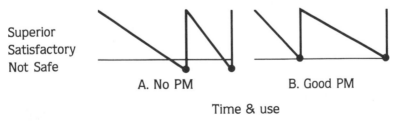

Superior
Satisfactory
Not Safe

A. No PM   B. Good PM

Time & use

Figure 2-1
PM to keep acceptable performance

Quality

For the same general reasons discussed above, good preventive maintenance helps to assure quality output. Tolerances are maintained within control limits. Naturally, productivity is improved and the investment in PM pays off with increased revenues.

Support to Users

If properly publicized, PM helps show equipment operators, production managers, and other equipment users that the maintenance function is striving to provide a high level of support. Note here that a PM program must be published so that everyone involved understands the value of PM, the investment required, and their own roles in the PM system.

Benefit/Cost

Too often, organizations consider only costs without recognizing the benefit and profits that are the real goal. PM allows a three-way balance between corrective maintenance, preventive maintenance, and production revenues.

DISADVANTAGES

In spite of all the good reasons for doing preventive maintenance, there are several potential problems that must be recognized and minimized.

Potential Damage

Every time a person touches a piece of equipment, there is potential for damage to occur through neglect, ignorance, abuse, or wrong procedures. Unfortunately, much high reliability equipment is serviced by low reliability people. The DC-10 engine pylon failure, the Three Mile Island nuclear power plant disaster, and many less-publicized accidents have been affected by inept preventive maintenance. Most of us have experienced car or home appliance problems that were caused by something that was done or not done at a previous service call. This situation gives rise to the slogan: if it works, don't fix it.

Infant Mortality

New parts and consumables have a higher probability of being defective, or failing, than exists with the materials that are already in use. Replacement parts are too often not subjected to the same quality assurance and reliability tests as parts that are put into new equipment.

Parts Use

Replacing parts at PM, rather than waiting until a failure occurs, will obviously terminate that part's useful life before failure and therefore requires more parts. This is part of the tradeoff between parts and labor and downtime, of which the cost of parts will usually be the smallest component. It must, however, be controlled.

Initial Costs

Given the time value of money and inflation that causes a dollar spent today to be worth more than a dollar spent or received tomorrow, it should be recognized that the investment in PM is made earlier than when those costs would be incurred if equipment were run until failure. Even though the cost will be incurred earlier, and may even be larger than corrective maintenance costs would be, the benefits in terms of equipment availability should be substantially greater from doing PM.

Access to Equipment

One of the major challenges when production is at a high rate is for maintenance to gain access to equipment in order to perform PM. This access will be required more frequently than it is with breakdown-driven maintenance. A good PM program requires the support of production, with immediate notification of any potential problems and willingness to coordinate equipment availability for inspections and necessary PMs.

The reasons for and against doing preventive maintenance are summarized in the following list. The disadvantages are most pronounced with fixed interval PM. On-condition and condition monitor PM both emphasize the positives and reduce the negatives.

Advantages
1. Performed when convenient
2. Increases equipment uptime
3. Maximum production revenue
4. Standardizes procedures, times, and costs
5. Minimizes parts inventory
6. Cuts overtime
7. Balances workload
8. Reduces need for standby equipment
9. Improves safety and pollution control
10. Facilitates packaging tasks and contracts
11. Schedules resources on hand
12. Stimulates preaction instead of reaction
13. Indicates support to user
14. Assures consistent quality
15. Promotes benefit/cost optimization

Disadvantages
1. Exposes equipment to possible damage
2. Failures in new parts
3. Uses more parts
4. Increases initial costs
5. Requires more frequent access to equipment

SUMMARY

PM can be both underdone and overdone. Most equipment failures result from too little PM. That is the most common situation since it is cheap and easy to do nothing. Intelligent preventive maintenance delivers benefits that should greatly exceed the costs.

CHAPTER 3

Designing a PM Program

FAILURE DATA

Valid failure data provide the intelligence for a PM program. After all, the objective of PM is to prevent those failures from recurring. A failure reporting system should identify the problem, cause, and corrective action for every maintenance repair work order or field service emergency call. An action group, prophetically called the Failure Review and Corrective Actions (FRACAS) Task Force, can be very effective for involving responsible organizations in both detailed identification of problems and causes, and assignment of both short- and long-term corrective action. The following are typical factory and field problems and codes:

A.	Not operable	O.	Other
B.	Below rate	P.	PM
I.	Intermittent	Q.	Quality
L.	Leak	S.	Safety
M.	Modification	W.	Weather
N.	Noise	Z.	No problem.

The following are typical cause codes:

1.	Not applicable	60.	Program
10.	Controls	70.	Materials
20.	Power	71.	Normal age wear
21.	External input power	72.	Damaged
22.	Main power supply	80.	Operator
30.	Motors	90.	Environment
40.	Drives	99.	Not found
50.	Transports	P.	PM.

The typical action codes are:

A.	Adjust/align	J.	Refurbish
B.	Calibrate	K.	Rebuild
C.	Consumables	L.	Lube
D.	Diagnose	M.	Modify
E.	Remove	P.	PM
F.	Remove and replace	R.	Repair
G.	Remove and reinstall	T.	Train
H.	Install	X.	Not complete
I.	Inspect	Z.	Not known.

These parameters and their codes should be established to fit the needs of the specific organization. For example, an organization with many pneumatic and optical instruments would have sticky dials and dirty optics that would not concern an electronically oriented organization. Note also that the code letters are the same, whenever possible, as the commonly used word's first letter. For example, leak = L, and noise = N. PMs are recorded simply as P/P/P. The cause codes, which may be more detailed, can use numbers and subsets of major groups, such as all power will be 20s, with external input power = 21, main power supply = 22, and so on.

It is possible, of course, to write out the complete words. However, analysis—whether done by computer or manually—requires standard terms. Short letter and number codes strike a balance that aids short reports and rapid data entry.

Use on the equipment at every failure should also be recorded. A key to condition monitoring PM is knowing how many hours, miles, gallons, activations, or other kind of use have occurred before an item failed. This requires hour meters and similar instrumentation on major equipment. Use on related equipments may often be determined by its relationship to the parent. For example, it may be determined that if a specific production line is operating for seven hours, then the input feeder operates five hours (5/7), the mixer two hours (2/7), and the packaging machine four hours (4/7).

It is also important to determine the valid relationship between the cause of the problem and the recording measurement. For example, failures of an automotive starter are directly related to the number of times the car engine is started and only indirectly to odometer miles. If startup or a particular activity stresses the equipment differently from normal use, those special activities should be recorded.

Figure 3-1 is a combination work order and completion form. This form is printed by the computer on plain paper with the details of the work order on the top, space in the center for labor and materials for work orders that take a day or less, and a completion blank at the bottom to show when the work was started, when it was completed, the problem/cause/action codes, and meter reading. Labor on work orders that take more than one day is added daily from time sheets and accumulated against the work order. Figure 3-2 shows the computer input screen for a simple service call report form that gathers minimum information necessary for field reporting. Those forms may be filed in a manual system by equipment identification number,

```
                         W O R K     O R D E R

ORDER#:   1926         PAD#: 45524        TYPE:  A              PRI:  8

  REQUESTED BY         DEPARTMENT      TELEPHONE#    TGT START       TGT COMPLETE
                                                   DATE    TIME      DATE    TIME
LARRY BEALE            MAINT PLAN                  5/ 6/82   830   5/ 6/82   1200
                                       EXT. 345

              DESCRIPTION                                    EQUIPMENT
A/C 44 PM-A.                                        ID:   44
                                                 NAME:   AIR COND
                                                  LOC:   BENDIX
                                                         CPTR RM 16


SPECIAL EQUIPMENT          ASSIGNED EMPLOYEE            PRECAUTIONS
CHARGER KIT          ID:   23456             PRD-PROD PERMT
                  NAME:   FELDMANN
                         HERBERT
DOC: A/C 544              ACCOUNTING: 123.555                      100%
---------------------------------------------------------------------------
                L A B O R    U S E D    (ONLY FOR SINGLE-DAY JOBS)

                         TOTAL HOURS-MINUTES   $    $    $    $
DATE      PERSON OR EQPT    WORK   TVL  DELAY  OT  MILES MEALS  PH  DIEM



                  M A T E R I A L    P O S T I N G
    DATE              PART#                DESC        QTY     EXTD  COST
  5/ 6/82           603552        KIT    ,FREON,A/C CHG  1        12.75




                              TOTAL MTL COST $
---------------------------------------------------------------------------
                         C O M P L E T I O N


                                               CODES    CURR
            DATE       TIME                     PBM:     METER
    STARTED:                                    CAU:     READ:
    COMPLETED:                                  ACT:



-------------------------------------------
SIGNATURE                   DATE
```

Figure 3-1
A combination work order and completion form.

```
            1 T   C L O S E   S E R V I C E   C A L L

CALL #:   2521   CUST ACCT #:  4543   NAME: KODAK-PARK CENTRAL FACILITIES
EMPL #:   2179          NAME: FELDMANN

STATUS (CMP=1)(CAN=2)(CNT=3): 1   ARR: 3/16/82 1030   CPLT: 3/17/82 1430

EQUIP ID#: C5      DESC: COMPRESSOR,10T,I-R           METER: 5250

HOURS-MINUTES                     $    $       $    $          CODES
WORK   TVL   DELAY   OT          MI   MEALS    PH   DIEM    PBM CAU ACT
8-50   1-10  1-00    -          4.00  7.50    .80           M   70  M

DESC: REPLACED WORN GEAR & INSTALLED STABILIZING BRACKET.

PART #           DESCRIPTION        UNIT CST   QTY   EXTD CST   RPR  RSP?
621112    GEAR  ,STNLS,HELI,18T     161.60      1     161.60     Y    Y
232240    OIL   ,GEAR,600W,QT          .66      6       3.96          Y
*               ,BRACKET,DWG 244    150.00      1     150.00          N

OTHER EQUIPMENT WORKED ON? N
TOTAL CALL:     HOURS        LABOR      MATERIALS      TOTAL
                11-00        206.25     315.56         521.81
```

Figure 3-2
Simple call report.

could be edge punched, or may be used as input for a computer system.

Figure 3-3 shows a complex service activity report that would be used for gathering data on expensive electronic equipment where the benefit from detailed information justifies the expense of collecting and analyzing the information.

FAILURES THAT CAN BE PREVENTED

Failure Modes, Effects, and Criticality Analysis (FMECA) provides a method for determining which failures can be prevented. Necessary inputs are the frequency of occurrence for each problem and cause combination and what happens if a failure occurs. Criticality of the failure is considered for establishing priority of effort. FMECA is a bottom-up approach that looks at every component in the equipment and asks, "Will it fail? And if so, how and why?" PM investigators are, of course, interested in how a component will fail so that the mechanism for failure can be reduced or eliminated. For example, heat is the most common cause of failure for electrical and mechanical components. Friction causes heat in assemblies moving relative to each other, often accompanied by material wear, and leads to many failures.

Any moving component is likely to fail at a relatively high rate and is a fine candidate for PM. The following are familiar causes of failure:

- Abrasion
- Abuse
- Bond Separation

- Operator Negligence
- Puncture
- Shelf life (due to

- Consumable Depletion
- Contamination
- Corrosion
- Dirt
- Fatigue
- Friction

other factors)
- Shock
- Stress
- Temperature Extremes
- Vibration
- Wear.

MAINTENANCE TO PREVENT FAILURES

Cleanliness is a watchword of preventive maintenance. Metal filings, fluids in the wrong places, ozone and other gases that deteriorate rubber components—all are capable of damaging equipment and causing it to fail. A machine shop, for example, that contains many electromechanical lathes, mills, grinders, and boring machines should have established procedures for assuring that the equipment is frequently cleaned and properly lubricated. In most plants, the best tactic is to assign responsibility for cleaning and lubrication to the machine's operator. There should be proper lubricants in grease guns and oil cans, and cleaning materials at every work station. Every operator should be trained on proper operator PM. A checklist should be kept on the equipment for the operator to initial every time the lubrication is done.

It is especially important that the lubrication be done cleanly. Grease fittings, for example, should be cleaned with waste material both before and after the grease gun is used. Grease attracts and holds particles of dirt. If the fittings are not clean, the grease gun could force contaminants between the moving parts, which is precisely what should be avoided. This is one example of how PM done badly can be worse than no PM at all.

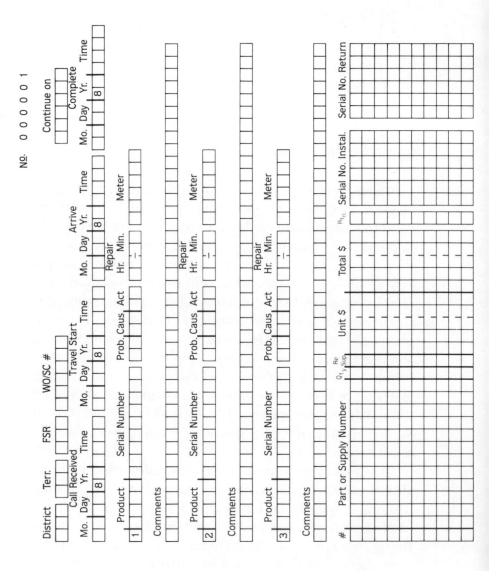

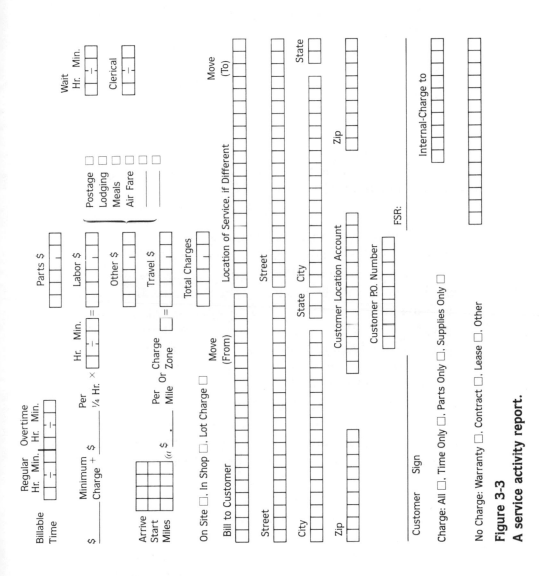

Figure 3-3
A service activity report.

PERSONNEL

Another tactic for assuring thorough lubrication is to have an "oiler" who can do all of the lubrication at the beginning of each shift. This may be better than having the operators do lubrication if the task is at all complicated or if the operators are not sufficiently skilled.

Whether or not operators will do their own lubrication, rather than an oiler, is determined by:

1. The complexity of the task
2. The motivation and ability of the operator
3. The extent of pending failures that might be detected by the Oiler but overlooked by operators.

If the lubrication can be properly done by operators, then it should be made a part of their total responsibility, just as any car driver will make sure that he has adequate gasoline in his vehicle. It is best if the operators are capable of doing their own PM. Like many tasks, PM should be delegated to the lowest possible level consistent with adequate knowledge and ability. If, however, there is a large risk that operators may cause damage through negligence, willful neglect, or lack of ability, then a maintenance specialist should do lubrication. The tasks should be clearly defined. Operators may be able to do some items while maintenance personnel will be required for others. Examples of how the work can be packaged will be described later.

PM tasks are often assigned to the newest maintenance trainee. In most cases, management is just asking for trouble if PM is regarded as low-status, undesirable work. If management believes in preventive maintenance, they should assign well-qualified personnel. Education and experience make a big difference in maintenance. Most organizations have at least one skilled maintenance person who can simply step onto the factory floor and sense—through sight, sound, smell, vibration, and temperature—the conditions in the factory. This person can tell in an instant that "The feeder on number 2 is hanging up a little this morning, so we'd better look at it." This person should be encouraged to take a walk around the factory at the beginning of every shift to sense what is going on and inspect any questionable events. The human senses of an experienced person are the best detection systems available today.

SERVICE TEAMS

A concept that is successfully applied in both factory and field service organizations is teams of three or four persons. This type of organization can be especially effective if equipment must have high uptime but requires lengthy maintenance time at failures or PMs. If individual technicians were assigned to the equipment, the person might well be busy on a lengthy project when a call comes to fix another machine. In an individual situation where a single person is responsible for specific machines, either the down machine would have to wait until the technician completes the first job and gets to the second, or if the second machine has greater priority, the first machine may be left inoperable. The technician then interrupts his task to take care of the second problem and must return later to complete the first, thus wasting time and effort. The optimum number of people can be calculated for any scenario, time, and effort. Figure 3-4 illustrates one situation in which two was the best team size.

A good technique for team work is to rotate the PM responsibility. The first week, Adam does all the PM, while Brad, Chuck, and Don do modifications and repairs. It may also help to assign Brad the short "do it now" (DIN) tasks for the same week. The next week, Brad does PM, and Don handles DIN, while Chuck and Adam attend to emergencies. Rotating PM has several advantages:

1. Responsibility is shared equally by all.

2. Doing a good PM job one week should reduce the breakdown emergency repairs in following weeks; thus a technician can benefit from the results of his own PM efforts.

3. Technicians' skills and interests vary, so that what one person fails to notice during his week will probably be picked up by another person the next week.

4. PM parts need be efficiently supplied to only the PM person.

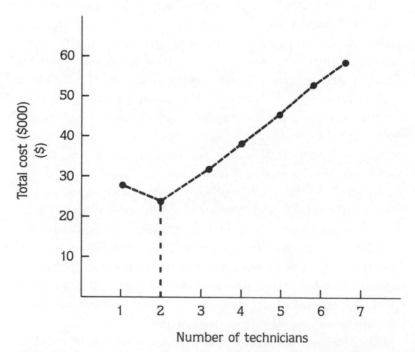

Number of technicians

Figure 3-4
Total maintenance costs for varied numbers of technicians.

WHEN TO START

The time to start is now. Don't let any more failures occur or information be lost. There is probably a lot of effort ahead, so get started now.

HOW TO START

The necessary items for establishing a PM system are:

1. Every equipment uniquely identified by prominent ID number or serial number and product type
2. Accurate equipment history records
3. Failure information by problem/cause/action
4. Experience data from similar equipment
5. Manufacturer's interval and procedure recommendations
6. Service manuals
7. Consumables and replaceable parts
8. Skilled personnel
9. Proper test instruments and tools
10. Clear instructions with a checklist to be signed off
11. User cooperation
12. Management support.

A typical initial challenge is to get proper documentation for all equipment. When a new building or plant is constructed, the architects and construction engineers should be required to provide complete documentation on all facilities and the equipment installed in them. Any major equipment that is installed after that should have complete documentation. Figure 3-5 is a checklist that should be given to anyone who purchases facilities and equipment that must be maintained. As can be seen, one of the items on this list is assuring availability of complete documentation and PM recommendations. Purchasing agents and facilities engineers are usually pleased to have such a checklist and will be cooperative if reminded occasionally about their major influence on life-cycle costs. This brings us back again to the principle of avoiding or minimizing the need for maintenance. Buying the right equipment in the beginning is the way to start. The best maintainability is eliminating the need for maintenance.

If you are in the captive service business or concerned with designing equipment that can be well maintained, you should recognize that the preceding has been aimed more at factory maintenance; but after all, that is an environment in which your equipment will often be used. It helps to view the PM system from the operator's and serviceperson's eyes to assure that everyone's needs are satisfied.

Figure 3-5
Maintenance considerations checklist for purchasing agents and facilities engineers.

	Yes	No	Comment
1. Standardization			
a. Is equipment already in use that provides the desired function?			
b. Is this the same as existing equipment?			
c. Are there problems with existing equipment?			
d. Can we maintain this equipment with existing personnel?			

	Yes	No	Comment

e. Are maintenance requirements com-
    patible with our current proce-
    dures?

2. Reliability and Maintainability
  a. Can vendor prove the equipment will
      operate at least to our minimum
      specifications? (Detail where pos-
      sible)
  b. Warranty of all parts and labor for 90
      days?
  c. Is design fault-tolerant?
  d. Are tests go/no go?

3. Spare Parts
  a. Is recommended spares list provided?
  b. Is the dollar total of spares less than
      10% of equipment cost?
  c. Do we already have usable spares?
  d. Can spares be purchased from other
      vendors?
  e. Are any especially high quality or ex-
      pensive parts required?

4. Training
  a. Is special technician training re-
      quired?
  b. Will manufacturer provide training?
      1. at no additional cost for first year?
      2. at our location as required?

	Yes	No	Comment

5. Documentation
    a. All tech manuals provided?
        1. installation
        2. operation
        3. corrective and preventive mainte-
           nance
        4. parts

6. Special Tools and Test Equipment
    a. Do we already have all required tools
        and test equipment?

    b. Can at least 95% of all faults be de-
        tected by use of proposed test
        equipment and procedures?

    c. Are calibration procedures minimum
        and clear?

7. Safety
    a. Are all UL/CSA, OSHA, EPA, and other
        applicable requirements met?

    b. Are any special precautions required?

    c. Can one person do all maintenance?

4

Economics of PM

BENEFITS VERSUS COSTS

PM is an investment. Like any thing in which we invest money and resources, we expect to receive benefits from PM that are greater than our investment. Targets for return on investment (ROI) vary, but in commerce and industry a project should have the potential of at least a 130 percent ROI. This means that you will gain 30 percent on your initial investment.

Some organizations also like to establish a maximum payback period. A few organizations want the investment at least to break even in the first year, while others allow a three-year period to achieve the ROI. The ROI is a reasonable financial control because we know that with inflation, money spent today on PM is going to be worth less per unit in the future when the benefit is reaped. We also know that forecasting the potential outcome is much more accurate in the short term than it is in the long term, which may be several years away.

The time effect is considered using present value techniques. Tables 4-1 thru 4-5 provide the factors necessary

Table 4-1   Future Value.
Single Payment Compound Amount Factor
$FV = P (1 + i)^n$

			INTEREST			
Periods	1%	2%	4%	10%	15%	20%
1	1.010	1.020	1.040	1.100	1.150	1.200
2	1.020	1.040	1.082	1.210	1.322	1.440
3	1.030	1.061	1.125	1.331	1.521	1.728
4	1.041	1.082	1.170	1.464	1.749	2.074
5	1.051	1.104	1.217	1.610	2.011	2.488
6	1.062	1.126	1.265	1.772	2.313	2.986
7	1.072	1.149	1.316	1.949	2.660	3.583
8	1.083	1.172	1.369	2.144	3.059	4.300
9	1.094	1.195	1.423	2.359	3.518	5.160
10	1.105	1.219	1.480	2.594	4.046	6.192
11	1.116	1.243	1.539	2.853	4.652	7.430
12	1.127	1.268	1.601	3.138	5.350	8.916
18	1.196	1.428	2.026	5.560	12.359	26.623
24	1.270	1.608	2.563	9.850		
36	1.431	2.040	4.104	30.913		
48	1.612	2.587	6.571			
60	1.817	3.281	10.520			

Table 4-2   Present Value.
Single Payment Present Worth Factor

$$PV = S \frac{1}{(1 + i)^n}$$

			INTEREST			
Periods	1%	2%	4%	10%	15%	20%
1	.990	.980	.962	.909	.870	.833
2	.980	.961	.925	.826	.756	.694
3	.971	.942	.889	.751	.658	.579
4	.961	.924	.855	.683	.572	.482
5	.951	.906	.822	.621	.497	.402
6	.942	.888	.790	.564	.432	.335
7	.933	.871	.760	.513	.376	.279
8	.923	.853	.731	.467	.327	.233
9	.914	.837	.703	.424	.284	.194
10	.905	.820	.676	.386	.247	.162
11	.896	.804	.650	.350	.215	.135
12	.887	.788	.625	.319	.187	.112
18	.836	.700	.494	.180	.081	.038
24	.788	.622	.390	.102	.035	.013
36	.699	.490	.244	.032		
48	.620	.387	.152			
60	.550	.305	.096			

**Table 4-3   Future Value of Annuity in Arrears.
Value of a Uniform Series of Payments**

$$USCA = P \frac{(1 + i)^n - 1}{i}$$

			INTEREST			
Periods	1%	2%	4%	10%	5%	20%
1	1.000	1.000	1.000	1.000	1.000	1.000
2	2.010	2.020	2.040	2.100	2.150	2.200
3	2.030	3.060	3.122	3.310	3.472	3.640
4	4.060	4.122	4.246	4.641	4.993	5.368
5	5.101	5.204	5.416	6.105	6.742	7.442
6	6.152	6.308	6.633	7.716	8.754	9.930
7	7.214	7.434	7.898	9.487	11.067	12.916
8	8.286	8.583	9.214	11.436	13.727	16.499
9	9.369	9.755	10.583	13.579	16.786	20.799
10	10.462	10.950	12.006	15.937	20.304	25.959
11	11.567	12.169	13.486	18.531	24.349	32.150
12	12.683	13.412	15.026	21.384	29.002	39.580
18	19.615	21.412	25.645	45.599	75.836	128.117
24	26.973	30.422	39.083	88.497	184.168	392.484
36	43.077	51.994	77.598	299.127	*	*
48	61.223	79.354	139.263	960.172	*	*
60	81.670	114.052	237.991	*	*	*

*Over 1,000

Table 4-4   Present Value of Annuity in Arrears.
Uniform Series Present Worth Factor

$$PVA_n = S \frac{(1 + i)^n - 1}{i(1 + i)^n}$$

Period	1%	2%	INTEREST 4%	10%	15%	20%
1	.990	.980	.962	.909	.870	.833
2	1.970	1.942	1.886	1.736	1.626	1.528
3	2.941	2.884	2.775	2.487	2.283	2.106
4	3.902	3.808	3.630	3.170	2.855	2.589
5	4.853	4.713	4.452	3.791	3.352	2.991
6	5.795	5.601	5.242	4.355	3.784	3.326
7	6.728	6.472	6.002	4.868	4.160	3.605
8	7.652	7.325	6.733	5.335	4.487	3.837
9	8.566	8.162	7.435	5.759	4.772	4.031
10	9.471	8.983	8.111	6.145	5.019	4.193
11	10.368	9.787	8.760	6.495	5.239	4.327
12	11.255	10.575	9.385	6.814	5.421	4.439
18	16.398	14.992	12.659	8.201	6.128	4.812
24	21.243	18.914	15.247	8.985	6.434	4.937
36	30.118	25.489	18.908	9.677	6.623	4.993
48	37.974	30.673	21.195	9.897	4.999	4.999
60	44.955	34.761	22.623	9967	6.665	5.000

Table 4-5   Capital Recovery.
Uniform Series with Present Value $1

$$CP = P \left(\frac{i(1 + i)^n}{(1 + i)^n - 1} \right)$$

			INTEREST			
Periods	1%	2%	4%	10%	15%	20%
1	1.010	1.020	1.040	1.100	1.150	1.200
2	.508	.515	.530	.576	.615	.654
3	.340	.347	.360	.402	.438	.475
4	.256	.263	.275	.315	.350	.386
5	.206	.212	.225	.264	.298	.334
6	.173	.179	.191	.230	.264	.301
7	.149	.155	.167	.205	.240	.277
8	.131	.137	.149	.187	.223	.261
9	.117	.122	.135	.174	.210	.248
10	.106	.111	.123	.163	.199	.239
11	.096	.102	.114	.154	.191	.231
12	.089	.095	.107	.147	.184	.225
18	.061	.067	.079	.120	.163	.208
24	.047	.053	.066	.111	.155	.203
36	.0033	.038	.051	.094	.151	.200
48	.026	.032	.045	.092	.150	.200
60	.022	.028	.043	.091	.150	.200

for evaluating how much an investment today must earn over the next three years in order to achieve our target return on investment. This calculation requires that we make a management judgment on what the inflation/interest rate will be for the payback time and what the pattern of those paybacks will be.

For example, if we spend $5,000 today to modify a machine in order to reduce breakdowns, the payback will come from improved production revenues, reduced maintenance labor, having the right parts, tools, and information to do the complete job and certainly less confusion.

The intention of this brief discussion of financial evaluation is to identify factors that should be considered and to recognize when to ask for help from accounting, control, and finance experts. Financial evaluation of PM is divided generally into either single transactions or multiple transactions. If payment or cost reductions are multiple, they may be either uniform or varied. Uniform series are the easiest to calculate. Nonuniform transactions are treated as single events that are then summed together. Tables 4-1 through 4-5 are done in periods of time and interest rates that are most applicable to maintenance and service managers. The small interest rates will normally be applicable to monthly events, such as 2 percent per month for twenty-four months. The larger interest rates are useful for annual calculations. The factors are shown only to three decimal places since the data available for calculation are rarely even that accurate. The intent is to provide practical, applicable factors that avoid "overkill." If more detailed factors are needed, or different time periods or interest rates, they can be found in most economics and finance texts. Note that future value factors (Tables 4-1 and 4-3) are larger than 1, as are present values for a stream of future payments (Table 4-4). This is because interest rates are added to the base number of 1, and will always result in a multiplier above 1. On the other hand, present value of a single future payment (Table 4-2) and capital recovery (Table 4-5 after the first year) result in factors of less than 1.000. The table factor is multiplied by the money involved to give the answer. Many programmable calculators can also work out these formulas. If, for example, interest rates are 15 percent per year and the total amount is to be repaid at the end of three years, refer to Table 4-1 on future value. Find the factor 1.521 at the intersection of three years and 15 percent. If our example cost is $35,000, it is multiplied by the factor to give: $35,000 × 1.521 = $53,235 due at the end of the term.

Present values from Table 4-2 are useful for determining how much we can afford to pay now to recover, say, $44,000 in expense reductions over the next two years. If the interest rates are expected to be under 20 percent, then $44,000 × .694 = $30,536. Note that a dollar today is worth more than a dollar received in the future.

The annuity tables are for uniform streams of either payments or recovery. Table 4-3 is used to determine the value of a uniform series of payments. If we start to save now for a future project that will start in three years, and save $800 per month through reduction of one person, and the cost of money is 2 percent per month, then $800 × 51.994 = $41,595 should be ready in your bank account at the end of thirty-six months. The factor 51.994 came from thirty-six periods at 2 percent. The first month's $800 earns interest for thirty-six months. The second month's savings earn for thirty-five months, and so on. The use of factors is much easier than using single payment tables and adding the amount for $800 earning interest for thirty-six periods ($1,632.00) plus $800 for thirty-five periods ($1,599.92) and continuing for thirty-four, thirty-three, and so on, through one.

If I sign a purchase order for new equipment to be

rented at $500 per month over five years at 1 percent per month, then $500 × 44.955 = $22,478. Note that five years is sixty months in the period column of Table 4-4. Capital recovery Table 4-5 gives the factors for uniform payments, such as mortgages or loans that repay both principal and interest. To repay $75,000 at 15 percent annual interest over five years: $75,000 × .298 = $22,350 annual payments. Note that over the five years, total payments will equal $111,750 (5 × $22,350), which includes the principal $75,000 plus interest of $36,750. Also note that a large difference is made by whether payments are due in advance or in arrears.

A maintenance service manager should understand enough about these factors to do rough calculations, and then get help from financial experts for fine tuning. Even more important than the techniques used is the confidence in the assumptions. Control and finance personnel should be educated in your activities so that they will know what items are sensitive and how accurate (or best-judgment) the inputs are, and will be able to support your operations.

TRADING PM FOR CM AND DOWNTIME

Figure 4-1 illustrates the relationships between PM, corrective maintenance (CM), and lost production revenues. The vertical scale is dollars. The horizontal scale is the percentage of total maintenance devoted to PM. The percentage of PM ranges from zero (no PM) at the lower left intersection to nearly 100 percent PM at the far right. Note that the curve does not go to 100 percent PM since experience shows that there will always be some failures that require corrective maintenance. Naturally, the more of any kind of maintenance that is done, the more it will cost to do those activities. The tradeoff, however, is that doing more PM should reduce both corrective maintenance and downtime costs. Note that the downtime cost in this illustration is greater than either PM or CM. Nuclear power-generating stations and many production lines have downtime costs in excess of $10,000 per hour. At that rate, the downtime cost far exceeds any amount of maintenance, labor, or even materials that we can apply to the job. It is obvious that the most important effort is to get the equipment back

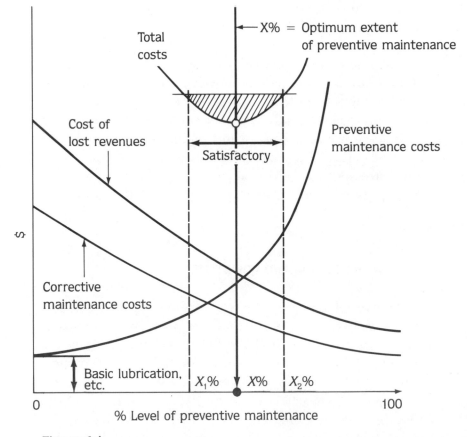

Figure 4-1
The relationship between cost and amount of preventive maintenance.

up without much concern for overtime or expense budget. Normally, as more PM is done, there will be fewer breakdowns and therefore lower corrective maintenance and downtime costs. The challenge is to find the optimum balance point.

As shown in Figure 4-1, it is better to operate in a satisfactory region than to try for a precise optimum point. Graphically, every point on the total-cost curve represents the sum of the PM costs plus CM costs plus lost revenues costs.

If you presently do no PM at all, then each dollar of PM effort will probably gain savings of at least ten dollars in reduced CM costs and increased revenues. As the curve shows, increasing the investment in PM will produce increasingly smaller returns as the break-even point is approached. The total-cost curve bottoms out and total costs begin to increase again beyond the break-even point. You may wish to experiment by going past the minimum-cost point some distance toward more PM. Even though costs are gradually increasing, subjective measures including reduced confusion, safety, and better management control that do not show easily in the cost calculations are still being gained with the increased PM. How do you track these costs? Figure 4-2 shows a simple record-keeping form that helps keep data on a month-by-month basis.

It should be obvious that you must keep cost data for all maintenance efforts in order to evaluate financially the cost and benefits of PM versus CM and revenues. A computerized maintenance information system is best, but data can be maintained by hand for smaller organizations. One should not expect immediate results and should anticipate some initial variation. This delay could be due to the momentum and resistance to change that is inherent in every electromechanical system, to delays in implementation through training and getting the word out to all personnel, to some personnel who continue to do things the old way, to statistical variations within any equipment and facility, and to data accuracy.

If you operate electromechanical equipment and presently have no PM, you are well advised to invest at least half of your maintenance budget for the next three months in preventive maintenance. You are probably thinking, "How do I put money into PM and still do the corrective maintenance?" The answer is that you can't spend the same money twice. At some point, you have to stand back and decide to invest in preventive maintenance

	Jan	Feb	Mar	Apr	May	Jun	Jly	Aug	Sep	Oct	Nov	Dec	Total
PM													
Labor	32	65	96	94	94	90	72						
Parts	23	38	49	56	68	65	54						
Total	55	103	145	150	162	155	126						
CM													
Labor	503	370	293	164	201	193	142						
Parts	231	213	181	185	199	196	157						
Total	734	583	474	349	400	389	299						
Lost revenues	407	397	320	290	330	320	362						
Total	1,196	1,083	939	789	82	864	787						

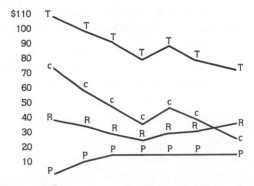

Figure 4-2
Maintenance PM, CM, and lost revenues costs, $00.

that will stop the large number of failures and redirect attention toward doing the job right once. This will probably cost more money initially as the investment is made. Like any other investment, the return is expected to be much greater than the initial cost.

One other point: it is useless to develop a good inspection and PM schedule if you don't have the people to carry out that maintenance when required.

Careful attention should be paid to the Mean Time to PM (MTPM). Many people are familiar with Mean Time to Repair (MTTR), which is also the Mean Corrective Time ($\bar{M}ct$). It is interesting that the term MTPM is not found in any textbooks the author has seen, or even in the author's own previous writings, although the term $\bar{M}pt$ is in use. It is easier simply to use Mean Corrective Time ($\bar{M}ct$) and Mean Preventive Time ($\bar{M}pt$).

$\bar{M}pt$ is calculated by PM Time/Number of PM events. That equation may be expressed in words as: the sum of all preventive maintenance time divided by the number of PM activities done during that time. If, for example, five oil changes and lube jobs on automobiles took 1.5, 1, 1.5, 2, and 1.5 hours, the total is 7.5 hours, which divided by the 5 events equals an average of 1.5 hours each. The statistics of PM are more thoroughly covered in the author's book, *Maintainability and Maintenance Management*. A few main points, however, should be emphasized here:

1. Mean Time Between Maintenance (MTBM) includes PM as well as CM.

2. Mean Maintenance Time is the weighted average of PM and CM and any other maintenance actions, including modifications and performance improvements.

3. Inherent Availability (A_i) considers only failure and $\bar{M}ct$. Achieved availability (A_a) adds in PM, though in a perfect support environment. Operational Availability (A_o) includes all actions in a realistic environment.

CHAPTER 5

Nondestructive Inspection

As was pointed out previously, inspection is a key to detecting the need for PM. It should be nondestructive so that it will not harm the equipment. Some common methods of nondestructive testing (NDT) are outlined below:

1. *Body Senses*
   —Sight
   —Smell
   —Sound
   —Taste
   —Touch
2. *Temperature*
   —Thermister
   —Thermometer
   —Crayons; stickers; paints
   —Infrared
   —Thermopile
   —Heat flow
3. *Vibration Wear*
   —Accelerometer

   —Stethoscope
   —Stroboscope
   —Ultrasonic listening
   —Laser alignment

4. *Materials Defects*
   —Magnetics
   —Penetrating dyes
   —Eddy currents
   —Radiographs
   —Ultrasonics
   —Rockwell hardness
   —Sonic resonance
   —Corona listener
   —Fiber optics bore scopes

5. *Deposits, Corrosion, and Erosion*
   —Ultrasonics
   —Radiographs
   —Cathodic potential
   —Weight
6. *Flow*
   —Neon freon detector
   —Smoke bomb
   —Gas sensor
   —Quick connect gauges
   —Manometer
7. *Electrical*
   —Cable fault detector
   —Outlet checker
   —Hipot
   —VOM
   —Oscilloscope
   —Static meter gun
   —Frequency recorder
   —Phase angle meter
   —Circuit-breaker tester
   —Transient voltage
8. *Tension*
   —V-belt depression
   —Backlash feeler
   —Torque meter
9. *Chemical/Physical*
   —Spectrographic oil analysis
   —Humidity
   —Water or antifreeze in gases and liquids
   —O_2
   —CO_2
   —pH
   —Viscosity
   —Metals present.

Tests may be made on functionally related components or on the output product. For example, most printing presses, copiers, and duplicators are intended to produce high-quality images on paper. Inspection of those output copies can show whether the process is working properly. Skips, smears, blurs, and wrinkles will show up on the copy. A good inspector can tell from a copy exactly what roll is wearing or what bearing is causing the skips. Careful inspection, which can be done without "tearing down" the machine, saves both technician time and exposure of the equipment to possible damage.

Rotating components find their own best relationship

to surrounding components. For example, piston rings in an engine or compressor cylinder quickly wear to the cylinder wall configuration. If they are removed for inspection, the chances are that they will not easily fit back into the same pattern. As a result, additional wear will occur and the rings will have to be replaced much sooner than if they were left intact and performance-tested for pressure produced and metal particles in the lubricating oil.

HUMAN SENSES

We humans have a great capability for sensing unusual sights, sounds, smells, tastes, vibrations, and touches. Efforts should be made by every maintenance manager to increase the sensitivity of his own and his personnel's human senses. Experience is generally the best teacher. Often, however, we experience things without knowing what we are experiencing. A few hours of training in what to look for could have high payoff.

Human senses are able to detect large differences but are generally not sensitive to small changes. Time tends to have a dulling effect. Have you ever tried to determine if one color was the same as another without having a sample of each to compare side by side? If you have, you will understand the need for standards. A standard is any example that can be compared to the existing situation as a measurement. Quantitative specifications, photographs, recordings, and actual samples should be provided. The critical parameters should be clearly marked on them with display as to what is good and what is bad. It is best if judgments can be reduced to "go/no go." Figure 5-1 shows such a standard.

As the reliability-based PM program develops, samples

should be collected that will help to pinpoint with maximum accuracy how much wear can take place before
problems will occur. A display where craftsmen gather can
be effective. A framed four-foot by four-foot pegboard
works well since shafts, bearings, gears, and other components can be easily wired to it or hung on hooks for display.
An effective, but little used, display area where notices can
be posted is above the urinal or on the inside of the toilet
stall door. Those are frequently viewed locations and allow
people to make dual use of their time.

SENSORS

Since humans are not continually alert or sensitive to
small changes, and cannot get inside small spaces, especially when operating, it is necessary to use sensors that
will measure conditions and transmit information to external indicators. Sensor technology is progressing rapidly;
there have been considerable improvements in capability,
accuracy, size, and cost. Pressure transducers, temperature
thermocouples, electrical ampmeters, revolution counters,
and a liquid height level float are examples found in most
automobiles. Accelerometers, eddy-current proximity sen-

Tip must protrude at least .500 cm

Figure 5-1
"Go/no go" standards.

sors and velocity seismic transducers are enabling the techniques of motion, position, and expansion analysis to be increasingly applied to large numbers of rotating equipments. Motors, turbines, compressors, jet engines, and generators can use vibration analysis. Figure 5-2 shows accelerometers placed on a rotating shaft. The accelerometers are usually permanently attached to equipment at two positions 90° apart, perpendicular to the rotating axes. Measurement of their output may be taken by portable test meters and chart recorders, or by permanently attached recorders, often with alarms that indicate when problem thresholds are exceeded. Such devices may automatically shut down equipment to prevent damage.

The normal pattern of operation, called its "signature," is established by measuring the performance of equipment under known good conditions. Comparisons are made at routine intervals, such as every thirty days, to determine if

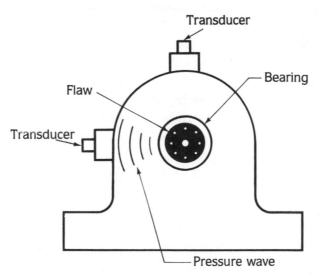

Figure 5-2
An accelerometer to measure the vibration of a rotating shaft.

any of the parameters are changing erratically, and further, what the effect of such changes may be.

The spectrometric oil analysis process, commonly referred to as "SOAP," is very useful for any mechanical moving device that uses oil for lubrication. It tests for presence of metals, water, glycol, fuel dilution, viscosity, and solid particles. Automotive engines, compressors, and turbines all benefit from oil analysis. Most major oil companies will provide this service if you purchase lubricants from them. Experience indicates that the typical result is that less oil is used and costs are reduced from what they were before using SOAP.

The major advantage of SOAP is early detection of component wear. Not only does SOAP evaluate when an oil is no longer lubricating properly and should be replaced, it also identifies and measures small quantities of metals that are wearing from the moving surfaces. The metallic elements found, and their quantity, can indicate what components are wearing and to what degree so that maintenance and overhaul can be carefully planned. For example, presence of chrome would indicate cylinder-head wear, phosphor bronze would probably be from the main bearings, and stainless steel would point toward lifters. Experience with particular equipment naturally leads to improved diagnosis. The Air Force and commercial air lines have been refining the techniques on jet aircraft for many years. They find that SOAP, together with bore scoping to look inside an engine and vibration analysis, enables them to do a very good job of predicting when maintenance should be done. The aircraft maintenance techniques that required complete teardown of propeller-driven aircraft every 1,000 hours, whether they needed it or not, are rapidly vanishing in that industry. Many manufacturing plants can gain improvements through the same maintenance techniques.

THRESHOLDS

Now that instrumentation is becoming available to measure equipment performance, it is still necessary to determine when that performance is "go" and when it is "no go." A human must establish the threshold point which can then be controlled by manual, semiautomatic, or automatic means. First, let's decide how the threshold is set and then discuss how to control it. To set the threshold, one must gather information on what measurements can exist while equipment is running safely and what the measurements were just prior to or at the time of failure. Equipment manufacturers, and especially their experienced field representatives, will be a good starting source of information. Most manufacturers will run equipment until failure in their laboratories as part of their tests to evaluate quality, reliability, maintainability, and maintenance procedures. Such data are necessary to determine under actual operating conditions how much stress can be put on a device before it will break. There are many devices, such as nuclear reactors and flying airplanes, that should not be taken to the breaking point under operating conditions, but they can be made to fail under secure test conditions so that knowledge can be used to keep them safe during actual use.

Once the breaking point is determined, a margin of safety should be added to account for variations in individual components, environments, and operating conditions. Depending on the severity of failure, that safety margin could be anywhere from one to three standard deviations before the average failure point. As Figure 5-3 shows, one standard deviation on each side of the mean will include 68 percent of all variations, two standard deviations includes 95 percent, and three standard deviations is 98.7

percent. Where our mission is to prevent failures, however, only the left half of the distribution is applicable. This single-sided distribution also shows that we are dealing with probabilities and risk.

The earlier the threshold is set and PM done, the greater is the assurance that it will be done prior to failure. For example, if the mean time between failures (MTBF) is 9,000 miles with a standard deviation of 1,750 miles, then almost 98 percent of the failures can be eliminated by doing perfect PM at 5,500 miles [9,000 − (2 × 1,750)]. Note the word "perfect," meaning that no new problems are injected. That also means, however, that costs will be higher than need be since components will be replaced before the end of their useful life, and more labor is required.

Once the threshold set point has been determined, it should be monitored to detect when it is exceeded. The investment in monitoring depends on the time period over which deterioration may occur, means of detection, and benefit value. Figure 5-4 illustrates the need for automatic monitoring.

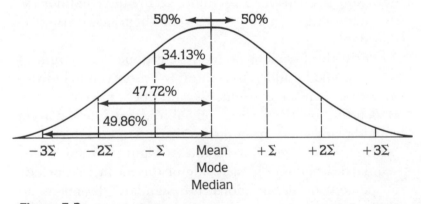

Figure 5-3
Normal distribution of failures.

If failure conditions build up quickly, the condition may not be easily detected by a human, and the relatively high cost of automatic instrumentation will be repaid.

The monitoring signal may be used to activate an annunciator that rings a bell or lights a red light. It may activate a feedback mechanism that reduces temperature or other parameters. A thermostat connected to a heating and air-conditioning system provides this feedback function to regulate temperature. The distinction between operational controls and maintenance controls is not important since the end result is reduced need for maintenance and notification that a problem is building up to a point where maintenance should be scheduled when convenient. A simple threshold indicator is the manometer shown in Figure 5-5.

This simple device can be effective in air conditioners, computer cabinets, office copiers, and any devices that rely on air flow. A spring loaded block can serve the same function in vacuum cleaners and other devices that must be moved and therefore can not rely on the pull/push of air against gravity. The purpose of a filter is to remove contaminant materials so that they will not clog coils, fans, electronic components, or optics. As the filter is doing its

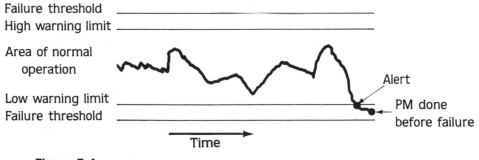

Figure 5-4
Control chart warning of possible failure before it occurs.

job, the caught contaminants reduce air flow. This will build to the point where equipment is straining to pull enough air and temperatures will probably begin to rise. At such a point, the filter should be changed or cleaned, which will restore equipment to normal operating conditions. This buildup of dirt can be easily detected by a difference in the air pressure. When the filter is operating efficienctly, air pushing on the entry side will be only slightly impeded and will have about the same pressure on the exit side. A small colored ball that fits inside the clear manometer tube will rest in the bottom when the air flow is balanced. As the filter becomes restricted, pressure on the entry will be greater than on the exit and the ball will be pushed to the exit side of the tube. Colored bands around the tube can indicate the threshold of safety versus a need to replace the filter. Since it will normally take at least several days and probably weeks for the filter to become

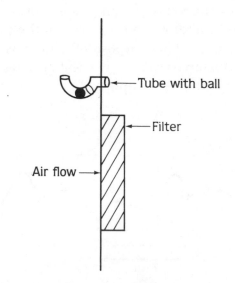

Tube with ball

Filter

Air flow

Figure 5-5
A simple manometer to warn of inadequate air flow.

clogged, checking of the manometer can be done on a routine inspection schedule and then maintenance can be performed as conditions require. This is certainly less expensive for both labor and materials than either routinely replacing the filter, whether it needs it or not, or letting it build up until equipment fails and both temperatures and tempers rise. More sophisticated sensors are certainly required where humans cannot or will not notice them, as well as remote communications and alarm systems.

The decision to put or not to put a filter in the air flow is a good example of initial investment in preventive maintenance that will pay off over the equipment life. Equipment would operate just fine initially without any filter and would, of course, cost less without those components. However, when contaminants build up on an electronic circuit board, coil or fan, extensive and expensive cleaning will have to be done to prevent the equipment from failing. Changing the filter is much easier than major equipment refurbishing, and the initial cost and replacement filters pay off through improved performance. As the automotive oil-filter advertising campaign said, "You can pay a little now, or a lot later."

CHAPTER 6

On-Condition Maintenance

Many of the topics introduced in the preceding chapter could also be applied to the subject of this chapter, "On-Condition Maintenance." Human factors have a major influence on maintenance. Very few people pay attention to the need of cost considerations, production priorities, or simple forgetfulness. For example, it is important to check the oil level in your car. Have you done it recently? If you have, you are one of a small minority. In the bye-gone days of full-service gasoline stations, the attendant's check under the hood often resulted in the sale of a quart of oil. Today, most people pump their own gas at a self-service station, and rarely check the engine oil level. In spite of improved engine designs that allow a car to run longer between oil changes, the number of engine overhauls that are required because of poor lubrication is increasing. An enterprising person could make good profits by checking oil for motorists at a self-service station and then selling them the needed replenishment oil.

In past years automobiles had instrument gages that showed the engine's variable analog conditions. But since the instruments were expensive and people paid little attention to them, economic and practical considerations led to the alternative use of so-called idiot lights. As explained earlier, people do not pay attention to slowly changing conditions, but the sudden appearance of a yellow or red light is usually immediately noticed and attended to.

The advantages of on-condition maintenance are not yet being fully exploited in commerce, government, or industry. While time-based intervals, such as every thirty days, are better than nothing, major additional improvements are possible if PM is done based on true on-condition need.

FAILURE PATTERNS

On-condition and condition monitor preventive maintenance require equipment failure patterns that give some warning before a disaster occurs. The aircraft industry has gathered extensive data that show patterns which should be similar to other commercial and industrial equipment. The "bath tub" curve shown in Figure 6-1 has long been

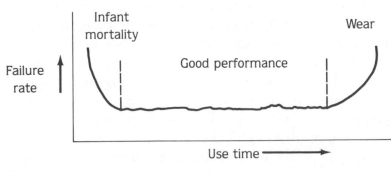

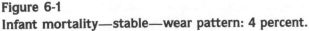

Figure 6-1
Infant mortality—stable—wear pattern: 4 percent.

regarded as the standard reliability failure plot. However, only 4 percent of aircraft components exhibit these characteristics of infant mortality, followed by a stabile period, followed by wear-out.

Electromechanical components such as relays and reciprocating engines are typical of this type of equipment and are best handled by on-condition and condition monitoring PM, with emphasis on inspection. The end-of-life wear failures can be alleviated by scheduled PM.

The largest group of aircraft components, about 68 percent of the items, follow a pattern like Figure 6-2. The failures occur during early use and can be weeded out by burn-in and good quality control. The best PM program for these is inspection and lubrication when needed.

The next most pronounced pattern is the flat failure rate, which occurs on about 14 percent of components, as shown in Figure 6-3. These items are definitely not candidates for fixed interval maintenance, with the exception of inspection and lubrication.

Figure 6-4 shows the 7 percent of components that fail because of operator and maintenance technician learning and human reliability problems. The failures are mostly caused by persons who push buttons and turn on equip-

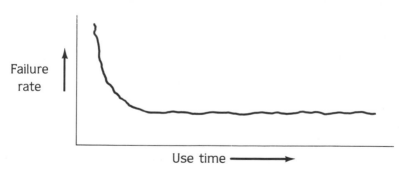

Figure 6-2
Early failures, then stabile life: 68 percent.

ment before they read instructions on machines that are not adequately idiot-proof.

Five percent have a failure pattern that trends higher over life, as shown in Figure 6-5.

Finally, the remaining 2 percent have few early failures and run for a long stabile life before wear sets in and increases failures. Figure 6-6 shows this rare pattern, which should be monitored to detect onset of the increasing failures, but should not otherwise be touched.

Of all components, only the 11 percent (Figures 6-1, 6-5, and 6-6) that have wear-out failures, are good candidates for PM done on a predetermined schedule, fixed

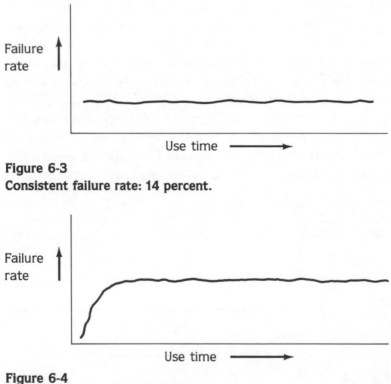

Figure 6-3
Consistent failure rate: 14 percent.

Figure 6-4
User-caused failures: 7 percent.

interval, periodic overhaul basis. The 89 percent majority is best handled by inspection, lubrication, and monitoring to track operating conditions, and then maintenance done when the quantitative criteria indicate a need.

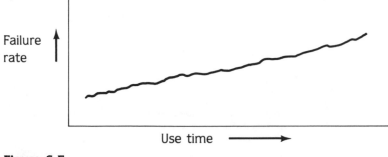

Figure 6-5
Increasing failures over life: 5 percent.

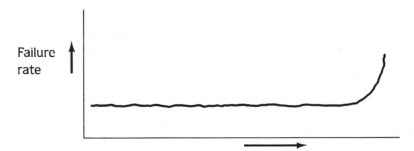

Figure 6-6
Wear after a good run life: 2 percent.

CHAPTER 7

Condition Monitoring Prediction

Condition monitoring requires an effective data-collection system that can be analyzed to provide information on when problems are building to a failure point. Since condition monitoring is statistically based, it is most effective where there are large numbers of similar equipments. Uses range from replacing stadium lights through automotive recall programs. Stadium lights, for example, will begin to burn out at about 450 hours of operation and nearly all the lights will burn out within the next 100 hours. The standard deviation is about 50 hours around an average of 500 hours. Replacing each light individually when it burns out would require an electrician, probably with a helper, to get the replacement bulbs, make the long and hazardous climb to the top of the light stand, and replace the bulb. This cannot be done safely at night. Also, once the climb has been to the top of the light stand, the electrician can easily replace all of the bulbs using little additional time.

Let us assume that the electrician and his helper costs thirty dollars per hour, and that it requires twenty minutes to get the tools, bulbs, keys, and ladders necessary to change even a single light bulb, ten minutes to climb each of twelve towers, and five minutes to change each of the fifteen bulbs in each tower. There are 180 (12 × 15) light bulbs. At the extreme of replacing them individually as they burn out (about once a year), each replacement would take thirty-five minutes and cost $17.50 ($30 × 35/60) in labor. That could amount to labor costs of $3,150 (180 × $17.50). At the other extreme, if all the lights were replaced at one time, labor cost would be only $370: [(20 minutes) + (12 towers × 10 minutes) + (120 lights × 5 minutes)] × ($30 cost/hour ÷ 60 minutes/hour). It becomes clear, thus, that the labor costs are less when all the lights are replaced at the same time.

But what about the costs of the bulbs themselves? Assume that each bulb costs $9.00 and will last 500 hours on the average, virtually none fail before 400 hours, and the distribution of failures is as shown in Figure 7-1. We also need to know how many can burn before the lighting is inadequate and they must be replaced. Since the lights are adjusted to overlap by about 50 percent, a practical criterion would be to replace the lights whenever two adjacent bulbs burn out. To illustrate: half the bulbs could be al-

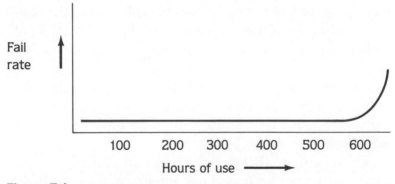

Figure 7-1
Typical distribution of light bulb failures.

lowed to burn out as long as they, like the black squares on a checkerboard, are not adjacent. That would be very unusual. An opposite extreme would be that only two lights burn out, but that they happen to be adjacent to each other.

For practical guidance, the electrician should check all the lights before he turns them off for the night to see how many have failed. That inspection will determine whether any replacements are necessary the next day. This is really on-condition maintenance.

The replacement criteria will usually result in about 450 hours of use on the average. That results in a 10 percent decrease in life (500 hours − 450 hours = 50 hours/500 = 10%). At nine dollars each, replacement bulbs for all 180 bulbs will cost \$1,620 (180 × \$9.00). The material cost for bulbs will increase by \$162 if they are all replaced at once.

Thus, the total cost of replacing all the bulbs at once is \$370 labor plus \$1,620 material for a total of \$1,990. Replacing the bulbs at individual failures, even assuming that we wait until two adjacent bulbs fail and then replace all that are out (about twenty individual changes), will cost \$190 more in getting materials (19 more occurrences of 20 minutes each at \$.50 per minute) and \$95 more in pole climbing (19 events × 10 minutes each × \$.50 per minute) for an increase of about \$285. That increase more than offsets the \$162 additional cost of bulbs. It can also be seen that the choice of maintenance method is affected by three variables: cost of bulbs, cost of labor per hour, and amount of time required.

The same principle—do all possible maintenance during a single access—has impact on other activities. For example, PMs, modifications, and low priority repairs may be held until a planned shutdown, or until the equipment fails. If maintenance is well planned, it should be possible to accomplish all the pending activities with little additional downtime. A manual or computerized record of all calls or work orders pending on equipment, or at a single location, is a valuable aid.

Another good idea is to inspect critical components routinely and to do any maintenance the inspection reveals is needed every time a service person is at the equipment. These techniques of routine interim maintenance (nicknamed TRIM) are especially valuable in situations where the wear leading to failure builds up slowly and the equipment is so reliable that it is attended to only infrequently.

FAILURE PATTERNS

Where many similar items of equipment are maintained, detection of failure patterns helps to stop failure modes before they spread to all pieces of equipment. An extreme example is where cracks are detected in an aircraft wing, every other similar aircraft should be inspected to determine if a failure mode exists, and if so, to eliminate it. The inspection and related data on wear versus use will provide valuable information to guide how much stress appears to be causing the defect, and when it is likely to become hazardous. In a factory with ten presses, a cambearing failure is cause for only minor concern. But if a second similar failure occurs, then all the remaining presses should be inspected and preparations made to correct the impending failure (assuming there is potential on all equipment) at a convenient time.

A problem/cause/action listing by frequency of occurrence can guide management toward priority areas that need analysis and preventive effort. Figure 7-2 shows a typical computer screen display using the codes and descriptions presented in Chapter 3.

```
6 E   P R O D U C T   B Y   P B M / C A U / A C T

ENTER PROD:  TRK     DESC:  TRUCK, 1/4T PICKUP  DATE:  1/ 1/82 THRU   3/31/82

      PROBLEM             CAUSE                  ACTION
COD FRQ  DESC       COD FRQ  DESC          COD FRQ  DESC

A   16  NOT OP      30  12  MOTORS         A   4  ADJUST/ALIGN
                                           C   4  CONSUMABLES
                                           H   3  REPAIR
                                           F   1  REMOVE/REPLACE
                    10   3  CONTROLS       A   2  ADJUST/ALIGN
                                           H   1  REPAIR
                    50   1  BRAKES         F   1  REMOVE/REPLACE

STRIKE NL TO CONT WITH NEXT MOST FREQUENT PROBLEM OR PREV-FLD FOR NEW PROD
```

Figure 7-2
A typical problem/cause/action report.

8

Scheduled Preventive Maintenance

When most people think of PM, they visualize scheduled, fixed interval maintenance that is done every month, every quarter, every season, or at some other predetermined intervals. That timing may be based on days, or on intervals such as miles, gallons, activations, or hours of use. The use of performance intervals is itself a step toward basing PM on actual need, instead of just on a generality.

The two main elements of fixed interval PM are procedure and discipline. Procedure means that the correct tasks are done and the right lubricants applied and consumables replaced at the best interval. Discipline requires that all the tasks are planned and controlled so that everything is done when it should be done. Both these areas deserve attention. The topic of procedures is covered in detail in following chapters.

Discipline is a major problem in many organizations. This is obvious when one considers the fact that many organizations do not have an established PM program. Further, organizations that do claim to have a program often fail to establish a good planning and control procedure to assure accomplishment. Elements of such a procedure include:

1. Listing of all equipment and the intervals at which it must receive PM

2. A master schedule for the year that breaks down tasks by month, week, and possibly even to the day

3. Assignment of responsible persons to do the work

4. Inspection by the responsible supervisor to make sure that quality work is done on time

5. Updating of records to show when the work was done and when the next PM is due

6. Follow-up as necessary to correct any discrepancies.

Note that there are variations within the general topic of scheduled fixed interval maintenance. Some tasks will be done every Monday whether or not they are necessary. Inspection may be done every Monday and PM tasks done if a need is indicated. Seasonal maintenance may be directed by environmental changes rather than by strict calendar date. Use meters, such as an automobile odometer, allow quantitative measure of use that can be related to the parameters that will need to be maintained. One must consider the relationship of components to the meter readings: for example, a truck's need for maintenance will vary greatly depending on whether it is used for long hauls or for local deliveries. A truck that is started every few miles and driven in stop-and-go, dusty city conditions will need more frequent mileage maintenance than the same truck used for long trips of continuous driving.

Seasonal equipment such as air conditioners, lawn mowers, salt spreaders, and snow blowers require special maintenance care at the end of each season in order to clean and refurbish them and store them carefully so that they will not deteriorate and will be ready for the next season. A lawn mower, for example, should have all gasoline drained from the tank and then be run until it stops because it has completely run out of fuel. This assures that gasoline is completely removed from the lines. Oil should be changed. The spark plug should be removed and cleaned. A tablespoon of engine oil should be poured into the cylinder through the spark plug hole and the cylinder pulled through several strokes to ensure that it is well lubricated. The spark plug should be put back in its hole loosely. Grass, dirt and other residue should be thoroughly cleaned from all parts of the mower. The blade should be sharpened and checked to see that it is in good balance. The mower should be stored in a dry place until it is needed again. Then, when the grass starts growing, all one has to do is fill the tank, tighten the spark plug, and connect the ignition wire. The motor should start on the second try. Careful preparation of equipment for storage will pay a major dividend when the equipment is needed in a hurry.

9

Lubrication

Friction of two materials moving relative to each other causes heat and wear. Great Britain has calculated that friction-related problems cost their industries over one billion dollars per annum. They coined a new term "tribology"—derived from the Greek work, "tribos," which means "rubbing"—to refer to new approaches to the old dilemma of friction, wear, and the need for lubrication. Technology intended to improve wear resistance of metal, plastics, and other surfaces in motion has greatly improved over recent years, but planning, scheduling, and control of the lubricating program is often reminiscent of a plant handyman wandering around with his long-spouted oil can.

Anything that is introduced onto or between moving surfaces in order to reduce friction is called a lubricant. Oils and greases are the most commonly used substances, although many other materials may be suitable. Other liquids and even gases are being used as lubricants. Air bearings, for example, are used in gyroscopes and other sensitive devices in which friction must be minimal. The functions of a lubricant are to:

1. separate moving materials from each other in order to prevent wear, scoring, and seizure
2. reduce heat
3. keep out contaminants
4. protect against corrosion
5. wash away worn materials.

Good lubrication requires two conditions: sound technical design for lubrication and a management program to assure that every item of equipment is properly lubricated.

LUBRICATION PROGRAM DEVELOPMENT

Information for developing lubrication specifications can come from four main sources:

1. Equipment manufacturers
2. Lubricant vendors
3. Other equipment users
4. Individuals' own experience.

Like most other preventive maintenance elements, initial guidance on lubrication should come from manufacturers. They should have extensive experience with their own equipment both in their test laboratories and in customer locations. They should know what parts wear and are frequently replaced. Therein lies a caution: a manufacturer could in fact make short-term profits by selling large numbers of spare parts to replace worn ones. Over the long term, however, that strategy will backfire and other vendors, whose equipment is less prone to wear and failure, will replace them.

Lubricant suppliers can be a valuable source of information. Most major oil companies will invest considerable time and effort in evaluating their customers' equipment to select the best lubricants and frequency or intervals for change. Figure 9-1 shows a typical report. Naturally, the vendor hopes that the consumer will purchase his lubricants, but the total result can be beneficial to everyone. Lubricant vendors perform a valuable service of communicating and applying knowledge gained from many users to their customers' specific problems and opportunities.

Experience gained under similar operating conditions by other users or in your own facilities can be one of the best teachers. Personnel, including operators and mechanics, have a major impact on lubrication programs. Table 9-1 shows typical codes for methods of lubrication, intervals, actions, and responsibility. Figure 9-2 shows a typical lubrication schedule. Detailing of specific lubricants and intervals will not be done here since they can be more effectively handled by the sources listed above.

The quality and the quantity of the lubricant applied are the two important conditions of any lube program. Lubrication properties must be carefully selected to meet the operating conditions. The viscosity of the oil (or the base oil, if grease is used) and additives to provide film strength under pressure are especially important for bearing lubrication.

Too little lubricant is usually worse than too much, but an excess can cause problems such as overheating and churning. The amount needed can range from a few drops per minute to a complete submersion bath.

A major step in developing the lubrication program is to assign specific responsibility and authority for the lubri-

LUBRICATION CHART
PREPARED BY
BEALE OIL COMPANY

NAME _SERVICE INFOSYSTEMS, INC._____ DATE _____2/10/82_____

EQUIPMENT			LUBRICANTS RECOMMENDED
ELECTRICAL DEPARTMENT			
Electric Motors			
Bearings	Ring Oiled	Ck W/Ch Y	MOBIL D.T.E. Oil Heavy Medium
Bearings	Hand Oiled	M	MOBIL D.T.E. Oil Heavy Medium
Bearings	Greased	2/Y	MOBILPLEX EP No. 1
Couplings	Greased	2/Y	MOBILPLEX EP No. 1
Motor—Generator Sets			
Bearings	Ring Oiled	Ck W/Ch Y	MOBIL D.T.E. Oil Heavy Medium
Bearings	Hand Oiled	M	MOBIL D.T.E. Oil Heavy Medium
Bearings	Greased	2/Y	MOBILPLEX EP No. 1
Couplings	Greased	2/Y	MOBILPLEX EP No. 1
York Refrigeration Compressor			
Brunner Refrigeration Compressor			
Crankcase & Cylinders	Splash		MOBIL D.T.E. Oil Heavy Medium
Transformers			
Circuit Breakers			
Compensators			
Insulating Oil			MOBILECT 25
Gear Motors			
Parallel Shafts	Splash	Ck M/Ch Y	MOBIL D.T.E. Oil Extra Heavy
Right Angle Worm Gears	Splash	Ck M/Ch Y	MOBIL 600 W Cylinder Oil
W — Weekly			
M — Monthly			
Y — Yearly			
2/Y — Twice Yearly			
Ck — Check			
Ch — Change			

Figure 9-1
Recommended lubricants.

cation program to a competent maintainability or maintenance engineer. The primary functions and steps involved in developing the program are to:

1. Identify every piece of equipment that requires lubrication

2. Assure that every major equipment is uniquely identified, preferably with a prominently displayed number

3. Assure that equipment records are complete for manufacturer and physical location

Table 9-1
Lubrication Codes.

Methods of application		Servicing actions	
ALS	Automatic lube system	CHG	Change
ALL	Air line lubricator	CL	Clean
BO	Bottle oilers	CK	Check
DF	Drip feed	DR	Drain
GC	Grease cups	INS	Inspect
GP	Grease packed	LUB	Lubricate
HA	Hand applied		
HO	Hand oiling	*Servicing intervals*	
ML	Mechanical lubricator	H	Hourly
MO	Mist oiler	D	Daily
OB	Oil bath	W	Weekly
OC	Oil circulation	M	Monthly
OR	Oil reservoir	Y	Yearly
PG	Pressure gun	NOP	When not operating
RO	Ring oiled	OP	OK to service when operating
SLD	Sealed		
SFC	Sight feed cups	*Service responsibility*	
SS	Splash system	MAE	Maintenance electricians
WFC	Wick feed oil cups	MAM	Maintenance mechanics
WP	Waste packed	MAT	Maintenance trades
		OPR	Operating personnel
		OIL	Oiler

4. Determine locations on each piece of equipment that needs to be lubricated

5. Identify lubricant to be used

6. Determine the best method of application

7. Establish the frequency or interval of lubrication

8. Determine if the equipment can be safely lubricated while operating, or if it must be shut down

```
CARBON PLANT AREA     LUBRICATION  SCHEDULE   WEEK OF 12/15/82      AREA CODE   17
SCHEDULE LUBE DATE                                                  PAGE        23
12/28/82  SHIFT 2

SEQUENCE   LUBE POINT                        TYPE OF             OIL
NUMBER     DESCRIPTION    METHOD  PLANTS  LUBRICATION  COMPLETED ADDED  CAPACITY

18-02      RACK & PINION   GUN      2      ANTI-SEIZE       Y N
18-03      SWIVEL          GUN      2      ANTI-SEIZE       Y N

************************* NO.20  ELEPHANT B 665-10 ***************************
20-01      BALL JOINT      GUN      2      ANTI-SEIZE       Y N
20-02      RACK & PINION   GUN      2      ANTI-SEIZE       Y N
20=03      SWIVEL          GUN      2      ANTI-SEIZE       Y N

************************* NO.21  ELEPHANT B-665-11 ***************************
22-01      BALL JOINT      GUN      2      ANTI-SEIZE       Y N
22-02      RACK & PINION   GUN      2      ANTI-SEIZE       Y N
22-03      SWIVEL          GUN      2      ANTI-SEIZE       Y N

                    - HOT PITCH STATION -

************************* PITCH PUMPS-3-INSIDE   ***************************
30-03      BEARINGS        GUN      4      ANTI-SEIZE       Y N

                    - 502 BUILDING -

************************* CONE CRUSHER-D-635-7   ***************************
38-00      CONE CRUSHER    RES      1        #5             Y N            35
38-02      BALL LOCK STOP  GUN      1        #1             Y N
38-03      BAND ROLL BRNGS GUN      4        #1             Y N

                    - 503 BLDG GRND FLOOR MERZ -

************************* 640 FARVAL AUTO LUBE SYS ***************************
40-01      PUMP            DRUM            TEMP 78          Y N
40-02      AIRLINE OILER   RES              140             Y N

                    - 504 BUILDING GROUND FLOOR -

************************* KENNEDY VAN SAUN BALL MIL ***************************
44-01      DRUM END BRNGS  GUN      4        1              Y N
44-02      GEAR DRIVE %N&S RES      2        10             Y N
```

Figure 9-2
Lubrication schedule.

9. Decide who should be responsible for any human involvement

10. Standardize lubrication methods

11. Package the above elements into a lubrication program

12. Establish storage and handling procedures

13. Evaluate new lubricants to take advantage of state of the art

14. Analyze any failures involving lubrication and initiate necessary corrective actions.

LUBRICATION PROGRAM IMPLEMENTATION

An individual supervisor in the maintenance department should be assigned the responsibility for implementation and continued operation of the lubrication program. This person's primary functions are to:

1. Establish lubrication service actions and schedules

2. Define the lubrication routes by building, area, and organization

3. Assign responsibilities to specific persons

4. Train lubricators

5. Assure supplies of proper lubricants through the storeroom

6. Establish feedback that assures completion of assigned lubrication and follows up on any discrepancies

7. Develop a manual or computerized lubrication scheduling and control system as part of the larger PM program

8. Motivate lubrication personnel to check equipment for other problems and to create work requests where feasible

9. Assure continued operation of the lubrication system.

It is important that a responsible person who recognizes the value of thorough lubrication be placed in charge. As with any activity, interest diminishes over time, equipment is modified without corresponding changes to the lubrication procedures, and state-of-the-art advances in lubricating technology may not be undertaken. A factory may have thousands of lubricating points that require attention. Lubrication is no less important to computer systems, even though they are often perceived as electronic. The computer field engineer must provide proper lubrication to printers, tape drives, and disks that spin at 3,600 rpm. A lot of maintenance time is invested in lubrication. The effect on production uptime can be measured nationally in billions of dollars.

10

Calibration

Calibration is a special form of preventive maintenance whose objective is to keep measurement and control instruments within specified limits. A "standard" must be used to calibrate the equipment. Standards are derived from parameters established by the National Bureau of Standards (NBS). Secondary standards that have been manufactured to close tolerances and set against the primary standard are available through many test and calibration laboratories and often in industrial and university tool rooms and research labs. Ohm meters, micrometers, scales, pumps, and flow meters are examples of equipment that should be calibrated at least once a year and before further use if subjected to sudden shock or stress.

STANDARDS

The government sets forth calibration system requirements in MIL-C-45662 and provides a good outline in the military standardization handbook MIL-HDBK-52, *Eval-*

uation of Contractor's Calibration System. The principles are equally applicable to any industrial or commercial situation. The purpose of a calibration system is to provide for the prevention of tool inaccuracy through prompt detection of deficiencies and timely application of corrective action. Every organization should prepare a written description of its calibration system. This description should cover the measuring of test equipment and standards, including:

1. Establishment of realistic calibration intervals
2. List all measurement standards
3. Established environmental conditions for calibration
4. Ensure the use of calibration procedures for all equipment and standards
5. Coordinate the calibration system with all users
6. Assure that equipment is frequently checked by periodic system or cross-checks in order to detect damage, inoperative instruments, erratic readings, and other performance degrading factors that cannot be anticipated or provided for by calibration intervals
7. Provide for timely and positive correction action
8. Establish decals, reject tags, and records for calibration labeling
9. Maintain formal records to assure proper controls.

INSPECTION INTERVALS

The checking interval may be in terms of time-hourly, weekly, monthly, or based on amount of use—every 5,000 parts, or every lot. For electrical test equipment, the power-on time may be the critical factor and can be measured through an electrical elapsed-time indicator.

Adherence to the checking schedule makes or breaks the system. The interval should be based on stability, purpose, and degree of usage. If initial records indicate that the equipment remains within the required accuracy for successive calibrations, then the intervals may be lengthened. On the other hand, if equipment requires frequent adjustment or repair, the intervals should be shortened. Any equipment that does not have specific calibration intervals should be (1) examined at least every six months, and (2) calibrated at intervals of no longer than one year. Adjustments or assignment of calibration intervals should be done in such a way that a minimum of 95 percent of equipment or standards of the same type is within tolerance when submitted for regularly scheduled recalibration. In other words, if more than 5 percent of a particular type of equipment is out of tolerance at the end of its interval, then the interval should be reduced until less than 5 percent is defective when checked.

CONTROL RECORDS

A record system should be kept on every instrument, including:

1. History of use
2. Accuracy
3. Present location
4. Calibration interval and when due
5. Calibration procedures and necessary controls
6. Actual values of latest calibration
7. History of maintenance and repairs

Figure 10-1 shows a typical calibration label.

Test equipment and measurement standards should be labeled to indicate the date of last calibration, by whom it was calibrated, and when the next calibration is due. When the size of the equipment limits the application of labels, an identifying code should be applied to reflect the serviceability and due date for next calibration. This provides a visual indication of the calibration serviceability status. A two-way check on calibration should be maintained by both the headquarters calibration organization and the instrument user. A simple means of doing this is to have a small form for each instrument with a calendar of weeks or months (depending on the interval required) across the top which can be punched and noticed to indicate the calibration due date. An example of this sort of form is shown in Figure 10-2.

If the forms are sorted every month, the cards for each instrument that should be recalled for check or calibration can easily be pulled out. The records and alert system can also be easily maintained by computer, as described in Chapter 15.

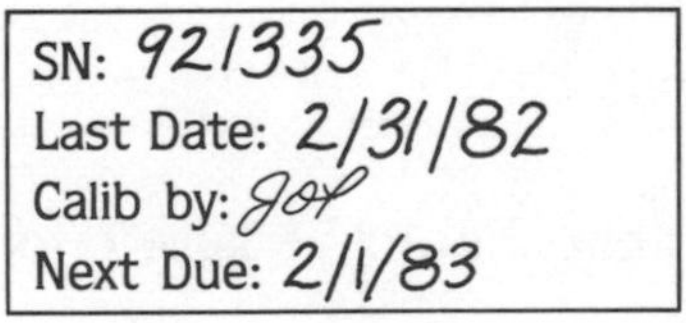

SN: *921335*
Last Date: *2/31/82*
Calib by: *Joe*
Next Due: *2/1/83*

Figure 10-1
A calibration label.

```
Month:   1   2   3   4   5 └──┘ 7   8   9   10   11 └──┘

SN: 921355                    Desc: Oscilloscope, Techtronix 213
User: Prototype Test Lab                    Acct: 121.355.722
      Bldg 32, Rm 13                        Int: 6 mo.
Attn: Mike Felluca        Tel: 334-9126
────────────────────────────────────────────────────────────
   Due — Date — Act            By         Comments
  12/ 1/80      12/ 4/80       JDP     OK
   6/ 1/81       6/15/81       HCF     OK
  12/ 1/82       8/ 3/82       JDP     Dropped. Repair/Recal.
  12/ 1/83
```

Figure 10-2
A calibration card.

Planning and Estimating

Planning is the heart of good inspection and preventive maintenance. As described earlier, the first thing to establish is what items must be maintained and what the best procedure is for performing that task. Establishing good procedures requires a good deal of time and talent. This can be a good activity for a new graduate engineer, perhaps as part of a training process that rotates him or her through various disciplines in a plant or field organization. This experience can be excellent training for a future design engineer.

Writing ability is an important qualification, along with pragmatic experience in maintenance practices. The language used should be clear and concise, using short sentences. Who, what, when, where, why, and how should be clearly described. A typical PM procedure is illustrated in Figure 11-1. The following points should be noted from this typical procedure:

1. Every procedure has an identifying number and title.
2. The purpose is outlined.
3. Tools, reference documents, and any parts are listed.

4. Safety and operating cautions are prominently displayed.

5. A location is clearly provided for the maintenance mechanic to indicate performance as either satisfactory or deficient. If deficient, details are written in the space provided at the bottom for planning further work.

The PM procedure may be printed on a reusable, plastic-covered card that can be pulled from the file, marked, and returned when the work order is complete, on a stan-

---

```
        WORK ORDER           CONTINUED         PAGE 3

PTRK1

        T R U C K    3 5 0 0    M I L E    O I L    C H A N G E

PURPOSE:  List cautions and steps required for changing oil.

REFERENCES:  Driver's manual for vehicle

CAUTIONS:  Assure vehicle is blocked securely before going under it!
           Hot oil from a recently operating motor can burn!
           Assure adequate ventilation when running gas or diesel engine!

PROCEDURES:
____   Get replacement oil from stockroom.
____   Get tools:  catch basin, oil spout, wrench, wipe.
____   Run motor at least 3 minutes to warm oil and mix contaminant particles.
____   Position vehicle on grease rack, lift, or oil change station.
____   Assure lift lock, blocks, and all safety devices are in safe position.
____   Position catch basin under oil drain.
____   Remove drain plug with wrench and drain oil into catch basin.
____   When oil slows to a trickle, replace drain plug.
____   If engine has a second sump, drain it the same way.
____   Open hood, remove oil fill cap, and fill engine with fresh oil.
____   Run engine 1 minute to circulate oil.  Check underneath for any leaks.
____   Check dip stick to assure oil level indicates in full area.
____   Clean any spilled oil.
____   Close hood and clean off any oil or finger prints.
____   Fill out oil change sticker with mileage and stick inside driver's door
       Hinge column.  Remove any old stickers.
____   Drive vehicle to parking area.  Be alert for indication of other problems.
____   Sign and date this checklist and write in mileage.


Completed by: _____  Date: _____

Vehicle ID#: _____  License: _____  Odometer miles: _____

Further work required:
```

Figure 11-1
A typical PM procedure.

dard preprinted form, or on a form that is uniquely printed by computer each time a related work order is prepared. Whatever the medium of the form, it should be given to the PM person together with the work order so that he has all the necessary information at his fingertips. The computer version has the advantage of single-point control that may be uniformly distributed to many locations. This makes it easy for an engineer at headquarters to prepare a new procedure or to make any changes directly on the computer and have them instantly available to any user in the latest version.

There are two slightly different philosophies for accomplishing the unscheduled actions that are necessary to repair defects found during inspection and preventive maintenance. One is to fix them on the spot. The other is to identify them clearly for later corrective action. This logic is outlined in Figure 11-2. If a "priority one" defect that could hurt a person or cause severe damage is observed, the equipment should be immediately stopped and "red tagged" so that it will not be used until repairs are made. Maintenance management should establish a guideline such as, "Fix anything that can be corrected within ten minutes, but if it will take longer, write a separate work request." The policy time limit should be set, based on:

1. Travel time to that work location
2. Effect on production
3. Need to keep the PM person on a precise time schedule.

Many small repairs can be effected quickest by the inspector who finds them. This avoids the need for someone else to travel to that location, reidentify the problem, and correct it. And it provides immediate customer satisfaction. More time-consuming repairs would disrupt the PM inspector's plans, which could cause other, even more serious problems to go undetected. The inspector is like a general practitioner who performs a physical exam and may give advice on proper diet and exercise, but who refers any problems he may find to a specialist.

The inspection or preventive maintenance procedure

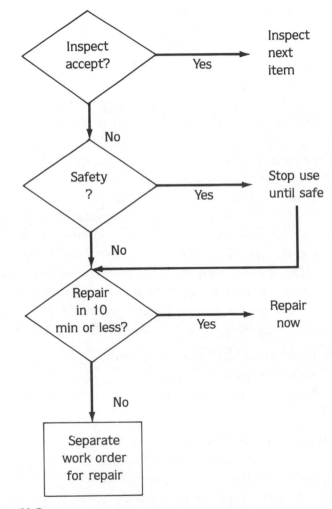

Figure 11-2
Logic for inspection findings.

form should have space where any additional action required can be indicated. When the procedure is completed and turned into maintenance control, the planner or scheduler should note any additional work required and see that it gets done according to priority.

ESTIMATING TIME

Since inspection or preventive maintenance is a standardized procedure with little variation, the tasks and time required can be accurately estimated. Methods of developing time estimates include consideration of such resources as:

1. Equipment manufacturers' recommendations
2. National standards such as *Chilton's* on automotive or *Means'* for facilities
3. Industrial engineering time-and-motion studies
4. Historical experience.

Experience is the best teacher, but it must be carefully critiqued to make sure that the "one best way" is being used and that the pace of work is reasonable.

The challenge in estimating is to plan a large percentage of the work (preferably at least 90 percent) so that the time constraints are challenging but achievable without a compromise in high quality. The tradeoff between reasonable time and quality requires continuous surveillance by experienced supervisors. Naturally, if a maintenance mechanic knows that his work is being time studied, he will follow every procedure specifically and will methodically check off each step of the procedure. When the industrial engineer goes away, the mechanic will do what he feels are necessary items, in an order that may or may not be satisfactory. As is discussed in Chapter 16, which deals with motivation, an experienced PM inspector mechanic can vary performance as much as 50 percent either way from standard, without most maintenance supervisors recognizing a problem or opportunity for improvement. Periodic checking against national or time-and-motion standards, as well as trend analysis of repetitive tasks, will help keep PM task times at a high level of effectiveness.

ESTIMATING LABOR COST

Cost estimates follow from time estimates simply by multiplying the hours required by the required labor rates. Beware of coordination problems where multiple crafts are involved. For example, one "Fortune 100" company has trade jurisdictions that require the following personnel in order to remove an electric motor: a tinsmith to remove the cover; an electrician to disconnect the electrical supply; a millwright to unbolt the mounts; and one or more laborers to remove the motor from its mount! That situation is fraught with inefficiency and high labor costs since all four trades must be scheduled together with at least three people watching while the fourth is at work. The cost will be at least four times what it could be, and is often greater if one of the trades does not show up on time. The best a scheduler can hope for is, if he has the latitude, to schedule the cover removal at say, 8:00 a.m., and the other functions at reasonable time intervals thereafter: electrician at 9:00, millwright at 10:00, and laborers at 11:00.

It is recommended that estimates be prepared on "pure" time. In other words, the exact hours and minutes that would be required under perfect scheduling conditions should be used. Likewise, it should be assumed that equipment will be immediately available from production, without delay. Delay time should be reported and scheduling problems should be identified so that they can be addressed separately from the hands-on procedure times. Note that people think in hours and minutes, so one hour and ten minutes is easier to understand than 1.17 hours.

ESTIMATING MATERIALS

Most parts and materials that are used for PM are well known and can be identified in advance. The quantity of each item planned should be multiplied by the cost of the item in inventory. The sum of those extended costs will be the material cost estimate. Consumables such as transmission oil should be enumerated as direct costs, but grease and other supplies used from bulk should be included in overhead costs.

FEEDBACK FROM ACTUAL

The time and cost required for every work order should be reported and analyzed to provide guidance for more accurate planning in future. It is important to determine what causes the task and times to change. Blindly assuming that the future will be like the past, or even that the past was done perfectly, may be an error. Comparisons should certainly be made between different individuals doing the same tasks in order to evaluate results in the amount of time required, what was accomplished during that time, quality of workmanship, and equipment performance as a result of their efforts.

Some people will argue that setting time standards for PM is counterproductive. They feel that the mechanic should be given as much time as he desires in order to ensure high-quality work. This is generally not true. In fact, the required tasks will generally expand or contract to fit the available time. Preventive maintenance inspection and lubrication can in fact be treated as a production operation with incentives for both time performance and equipment uptime capability. The standard maintenance estimating and scheduling techniques of time slotting, use of ranges, and calculations based on the log-normal distribution may be followed as reliable data and analytical competence are established. Since PM time and costs will typically comprise 30 to 60 percent of the maintenance budget, accurate planning, estimating, and scheduling are crucial to holding costs and improving profits.

CHAPTER 12

Shutdown Planning

Many operations halt for a period of time in order to consolidate all maintenance activities, accomplish inspections that cannot be done during operation, and accommodate seasonal raw material or personnel situations. In many industries it is very difficult for maintenance personnel to get their hands on equipment while demand for production is high. In food processing, for example, while "green pack" is underway, the freshly picked crops must be processed quickly before they spoil. Farmers' trucks are probably lined up at the processor's plant, which runs around the clock every day until the harvest is processed. The industry typically pays little attention to maintenance during this period. Then, when fresh pack is finished, all the accumulated problems will be attended to and the equipment is often completely overhauled.

The epitome of inspections and preventive maintenance done at shutdown is a nuclear-power-generating station. Even a relatively small 500-megawatt plant produces over $250,000 worth of electricity every day. When such a station is not generating electricity, that power must be purchased from other sources. With that revenue value exceeding $10,000 an hour, considerable attention should be paid to planning ahead so that the shutdown can be quickly accomplished and the equipment brought on line again in a minimal amount of time. Most power stations have shutdown planning specialists who devote their attention to planning, coordination, communication, and control of the thousands of activities that must be accomplished in order to assure continued safe operation.

CRITICAL PATH

Critical Path Method (CPM) planning is most effective for such shutdowns, just as it is for planning construction and new product development. If many tasks are involved, such as the thousands required for a nuclear plant, a computerized planning and tracking program is very helpful. A sample printout from such a program is shown in Figure 12-1.

A large wall chart is useful for planning and communication. Magnetic tabs or sliding-track tabs on which the activity number and a brief description may be written aid the process. If the task tabs are scaled to represent the number of hours or days that activity will require, then the tabs can be created and easily manipulated on the board in order to develop the best combination. As the activities are taking place, a vertical time line represented by a colored string can be moved from day to day to draw people's attention quickly to the status. Somewhat the same function in simpler Gantt Chart format can be provided by sheets of plexiglass fastened over gridded paper on the wall. Thin black drafting tape can be used to make the grids and the plexiglass can be written on with wipeable marking pens. In addition to their function as a planning tool, these wall boards allow all involved persons to see the status of all work quickly and provide a visual tool for scheduling balanced work loads. The Australians call these "stagger boards" since they are useful for staggering the workload. It is possible to show much the same information on computer terminals, but the screen switching and level of sophistication presently involved are too complex for most maintenance planners. The combination of

```
GROUP..PIPEFITTERS              ROCHESTER GAS & ELECTRIC          DATE..04/23/81
                                     GINNA STATION                DAY........THU
                                OUTAGE SHIFT WORK LIST            SHIFT......DAY
                                                                 PAGE........01

                      ***SCHEDULE***              ***FEEDBACK***
|ACTIV|                      |         |ACT|ACT|*PLEASE INCLUDE ANY EXTRA WORK*
| NO. |    DESCRIPTION       | RESOURCE|MEN|HRS|IF OFF SCHEDULE, PLEASE COMMENT
================================================================================
|11047| INSTALL CREVICE      |         |   |   |    COMPLETE      YES  NO       |
|     | EQUIPMENT IN CV      |         |   |   |    ON SCHEDULE YES  NO       |
|     | AND TURB. BLDG.      |         |   |   |                             |
|     | 49-0-0-1-183-0/0-695|         |   |   |                             |
================================================================================
|11277| HOOK-UP CC EQUIP TO  |         |   |   |    COMPLETE      YES  NO       |
|     | CC HANDHOLE AND TEST|         |   |   |    ON SCHEDULE YES  NO       |
|     | SYSTEM              |         |   |   |                             |
|     |                     |         |   |   |                             |
================================================================================
|13055| A DIESEL GENERATOR   |         |   |   |    COMPLETE      YES  NO       |
|     | AI&O                |         |   |   |    ON SCHEDULE YES  NO       |
|     |                     |         |   |   |                             |
|     | 49-0-13-1-531-1365  |         |   |   |                             |
================================================================================
|19018| PNEUMATIC            |         |   |   |    COMPLETE      YES  NO       |
|     | SOLIDIFICATION SYS  | ADDED JO|   |   |    ON SCHEDULE YES  NO       |
|     |                     | SWP     |   |   |                             |
|     | 49-0-13-1-530-1363  |         |   |   |                             |
================================================================================
|20001| SECONDARY SIDE VALVE|         |   |   |    COMPLETE      YES  NO       |
|     | PACKING             |         |   |   |    ON SCHEDULE YES  NO       |
|     |                     |         |   |   |                             |
|     | 49-0-13-1-530-1363  |         |   |   |                             |
================================================================================
|21030| B HOTWELL SAMPLE     |         |   |   |    COMPLETE      YES  NO       |
|     | PUMP                |         |   |   |    ON SCHEDULE YES  NO       |
|     |                     |         |   |   |                             |
|     |                     |         |   |   |                             |
================================================================================
|29001| `A` INSTRUMENT AIR   |         |   |   |    COMPLETE      YES  NO       |
|     | MAJOR OVERHAUL      |         |   |   |    ON SCHEDULE YES NO        |
|     |                     |         |   |   |                             |
|     | 49-0-13-1-531-1364  |         |   |   |                             |
================================================================================
```

Figure 12-1
Critical path control list.

schedule boards and the computer is the best kind of aid for most operators.

An important part of planning for inspections and preventive maintenance is learning from previous experience. Records should be made of what was done on every important work order. The problem/cause/action reporting should be analyzed for trends and the common failure items carefully inspected with the objective of eliminating future problems. Shutdown provides the opportunity for modifications that will reduce the need for future maintenance.

COORDINATION

Coordination with production is especially important for shutdown planning. The shutdown may be scheduled by production or it may be requested by maintenance. When such a shutdown is anticipated, the maintenance planner should initially gather information on all modifications that are pending and on any changes that should be consolidated and done while the equipment is available. An estimate of the labor, including crafts, number of people, and total time required, and of all parts and materials that will be needed should be prepared. The parts should be ordered so that they will arrive before the shutdown and gathered to specific areas of the stockroom when they do arrive. Bins or pallets should be clearly marked with the work order number and a bill of materials list so that the parts can be easily inventoried and it is obvious that they are there for a specific purpose and should not be borrowed. The maintenance planner should schedule required personnel to be ready when the equipment becomes available.

As each task is completed, or if a delay is encountered that will cause a due date to be missed, the project coordinator should be notified. On high-priority, complex projects such as shutdown at a nuclear plant or petrochemical refinery, a control team will manage all details and may hold short coordination meetings as often as once a shift, around the clock. When the installation is back on line, a summary analysis should be completed and written down as guidance for the next shutdown. Detailed records should be kept on what was done to every major equipment, particularly on what parts were replaced. Nondestructive test results help determine the expected life of components and allow better planning for both on-condition and condition monitoring.

13

Scheduling

Scheduling is, of course, one of the advantages to doing preventive maintenance over waiting until equipment fails and then doing emergency repairs. Like many other activities, the watchword should be "PADA," which stands for "Plan a Day Ahead." In fact, the planning for inspections and PM can be done days, weeks, and even months in advance in order to assure that the most convenient time for production is chosen, that maintenance parts and materials are available, and that the maintenance workload is relatively uniform.

Scheduling is primarily concerned with balancing demand and supply. Demand comes from the equipment's need for PM. Supply is the availability of the equipment, craftspeople, and materials that are necessary to do the work. Establishing the demand has been partially covered in the chapters on on-condition, condition monitoring, and fixed interval PM. Those techniques identify individual equipment as candidates for PM.

PRIORITIZING

When the individual pieces of equipment have been identified for PM, there must be a procedure for identifying the order in which they are to be done. Not everything can be done first. First In First Out (FIFO) is one way of scheduling demand. Using FIFO means that the next PM picked off the work request list, or the next card pulled from the file, is the next PM work order. The problem with this "first come, first served" method is that the more desirable work in friendly locations tends to get done while other equipment somehow never gets its PM. The improved method is Priority = Need Urgency × Customer Rank × Equipment Criticality. The acronym NUCREC will help in remembering the crucial factors.

NUCREC improves upon the Ranking Index for Maintenance Expenditures (RIME) in several ways:

1. The customer rank is added.
2. The most important item is given the number-one rating.
3. The number of ratings in the scale may be varied according to the needs of the particular organization.
4. Part essentiality may be considered.

A rating system of numbers 1 through 4 is recommended. Since most humans think of number 1 as the first priority to get done, the NUCREC system does number 1 first.

Need urgency ratings include:

1. Emergency; safety hazard with potential further damage if not corrected immediately; call back for unsatisfactory prior work
2. Downtime; facility or equipment is not producing revenue
3. Routine and preventive maintenance
4. As convenient, cosmetic.

The customer ranks are usually:

1. Top management

2. Production line with direct revenue implications

3. Middle management, research and development facilities, frequent customers

4. All others.

The equipment criticality ratings are:

1. Utilities and safety systems with large area effect
2. Key equipment or facility with no backup
3. Most with impact on morale and productivity
4. Low, little use or effect on output.

The product of the ratings gives the total priority. That number will range from 1 (which is $1 \times 1 \times 1$) to 64 ($4 \times 4 \times 4$). The lowest number work will be first priority. A "1" priority is a first-class emergency. When several work requests have the same priority, labor and materials availability, locations, and scheduling fit may guide which is to be done first.

The priorities should be set in a formal meeting of production and maintenance management at which the equipment criticality number is assigned to every piece of equipment. Similarly, a rank number should be applied to every customer and the need urgency should be agreed on. With these predetermined evaluations, it is easy to establish the priority for a work order either manually by taking the numbers from the equipment card and the customer list and multiplying them by the urgency, or by having the computer do so automatically. Naturally, there may be a few situations in which the planner's judgment should override and establish a different number, usually a lower number so that the work gets done faster.

Ratings may rise with time. A good way to assure that PM gets done is to increase the need urgency every week. In a computer system that starts with PM at 3, a PM that is to be done every month or less frequently can be elevated after one week to a 2, and finally to a 1 rating. Those increases should assure that the PM is done within a reasonable amount of time. If preventive maintenance is required more often, the incrementing could be done more rapidly.

Dispatch of the PM work orders should be based on the demand ordered by priority, consistent with availability of labor and materials. As discussed earlier, PM provides a good buffer activity in service work since time within a few days is not normally critical. The NUCREC priority system helps assure that the most important items are done first.

Some pressure will be encountered from production people who want a particular work request filled right away instead of at the proper time in the priority sequence. It can be helpful to limit the "criticality 1" equipments and "rank 1" customers to 10 percent, since, according to Pareto's Principle of the Critical Few, they will probably account for the majority of activity. If rank 2 is the next 20 percent, rank 3 is 30 percent, and the balance is 40 percent for rank 4, the workload should be reasonably balanced. If temporary work needs exist for selected equipment or a customer needs to be given a higher priority, then one equipment should be moved to a lower criticality for each equipment that is moved higher. After all, one objective of prioritization is to assure that work gets done in proper sequence. A preventive maintenance action that is done on time should assure that equipment keeps operating and that emergency work is not necessary.

COORDINATION WITH PRODUCTION

Equipment is not always available for preventive maintenance just when the maintenance schedulers would like it to be. An overriding influence on coordination should be a cooperative attitude between production and maintenance. This is best achieved by a meeting between the maintenance manager and production management, including the foreman level, so that what will be done to prevent failures, how this will be accomplished, and what production should expect to gain in uptime may all be explained.

The cooperation of the individual machine operators is of prime importance. They are on the spot and most able to detect unusual events that may indicate equipment malfunctions. Once an attitude of general cooperation is established, coordination should be refined to monthly, weekly, daily, and possibly even hourly schedules. Major shutdowns and holidays should be carefully planned so any work that requires "cold" shutdown can be done during those periods. Maintenance will often find that they must do this kind of work on weekends and holidays, when other persons are on vacation.

Normal maintenance should be coordinated according to the following considerations:

1. Maintenance should publish a list of all equipment that is needed for inspections, preventive maintenance, and modifications; and the amount of cycle time that such equipment will be required from production.

2. A maintenance planner should negotiate the schedule with production planning so that a balanced work load is available each week.

3. By Wednesday of each week, the schedule for the following week should be negotiated and posted where it is available to all concerned; it should be broken down by days.

4. By the end of the day before the PM is scheduled, the maintenance person who will do the PM should have

seen the first-line production supervisor in charge of the equipment to establish a specific time for the PM.

5. The PMer should make every effort to do the job according to schedule.

6. As soon as the work is complete, the maintenance person should notify the production supervisor so that the equipment may be put back into use.

Overdue work should be tracked and brought up to date. PM scheduling should make sure that the interval is maintained between PM actions. For example, if a PM for May is done on the thirtieth of the month, the next monthly PM should be done during the last week of June. It is foolish to do a PM on May 30th and another June 1st, just to be able to say one was done each month. In the case of PM, the important thing is not the score but how the game was played.

OPPORTUNITY PM

It is often helpful to do PM when equipment suddenly becomes available, which may not be on a regular schedule. One method called techniques of routine interim maintenance (TRIM) was covered in Chapter 11. TRIM means generally that specified cleaning, inspection, lubrication, and adjustments are done at every service call. TRIM can be very effective.

Another variation is to convert (or expand) a repair call to include PM. A good work order or service call system will quickly show any PM, modification, or other work due when equipment work is requested. The system should also check parts availability and print pick lists. Parts required for PM replacement can then be taken to the site and all work done at one time. Unless production is in a hurry to use the equipment again as soon as possible, doing all work on an equipment during the single access is much more efficient than having to gain access several times in order to perform a few tasks each time.

ASSURING COMPLETION

A paper work order form is desirable for every inspection and PM job. If the work is at all detailed, a checklist should be used. The completed checklist should be returned to the maintenance office on completion of the work. Any open PM work orders should be kept on report until the supervisor has checked the results for quality assurance and signed off approval.

14

Record Keeping

The foundation records for preventive maintenance are the equipment files. In a small operation with under 200 pieces of complex equipment, the records can easily be maintained on paper. Two proven card-based systems are illustrated in Figures 14-1 and 14-2.

The equipment records provide information for purposes other than PM. The essential items include:

—Equipment identification number
—Equipment name
—Equipment product/group/class
—Location
—Use meter reading
—PM interval(s)
—Use per day
—Last PM due
—Next PM due
—Cycle time for PM
—Crafts required, number of persons, and time for each
—Parts required.

Figure 14-3 shows the equipment screen from the COmputerized Maintenance Management System (COMMS). The data elements can be seen in an easy-to-enter format with a display of three levels of PM (denoted in this figure as A, B, and C).

WORK ORDERS

All work done on equipment should be recorded on the equipment record or on related work order records that can be searched by equipment. The equipment failure and repair history provide vital information for analysis to determine if PM is effective.

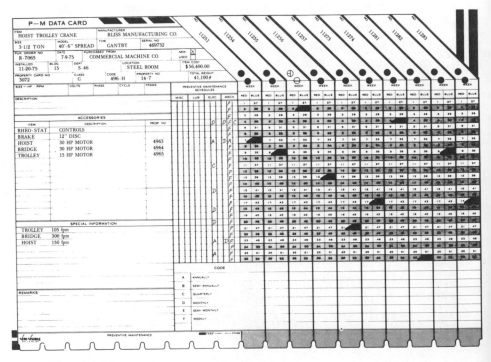

Figure 14-1
The Acme Veri-Visible® PM system.

How much detail should be retained on each record must be individually determined for each situation. Certainly, replacement of main bearings, crank shafts, rotors, and similar long-life items that are infrequently replaced should be recorded. That knowledge is helpful for planning major overhauls both to determine what has recently been done, and therefore should not need to be done at this event, and also for obtaining parts that probably should be replaced. There is certainly no need to itemize every nut, bolt, and light bulb.

The equipment records can be stored in a hanging file, if they are in paper form. Modern PM programs should be on, or preparing for, computerized records, as described in the next chapter.

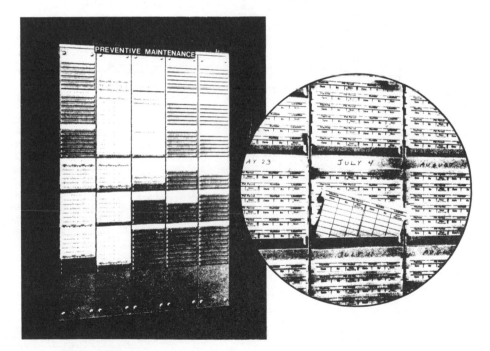

Figure 14-2
The Methods Research T-Card® system.

```
                    2 A   N E W   E Q U I P M E N T

ID: 123456789      DESC: TRUCK,1/4TN,4WD,DODGE      CRIT: 2
SN: D0008137       PROD: TRK   MFGR: DODGE   ACQD: 12/15/80   COST$ 9,600.  COND: N
INST: 12/15/80     WRTY TO 12/15/81   CONT#: C231   PRD$/HR:      45 MTR:  50 MI
USER ACCT:  58     NAME: PLT 2, GENL MAINT   ORGN: 125     EMPL:179300434
LOC: 3599 W. RIDGE, BAY 4   PRECAU 200-DISP AUTH
                     PM-A              PM-B              PM-C
INTV:           7500    MI        15000    MI        180    DA
USE/DAY:        65.0    MI           65    MI        1.0    DA
CY TI:               1.5               3.0               1.5
CR-HR-#:    GENM 1.5  1        GENM 3.0  1        GENM 1.5  1
                                I&E  .5  1

DOC:             M22               M23               M24
LAST:       12/15/80          12/15/80          12/15/80
NEXT:        6/ 1/81          11/17/81           6/14/81
                 PM-A              PM-A              PM-C
                                   PM-B
P1234567890A   DESC  FILTER,OIL,TRK,CRYSLER        1
8888                 OIL,10W-40,HD            6    6
```

Figure 14-3
The COMMS equipment data screen.

CHAPTER 15

Computer Assistance

Modern computers offer the capability to maintain accurate records, update them instantly with the most recent information, and do accurate calculations far more rapidly than a human can. Elements of the COmputerized Maintenance Management System (COMMS)®, a proprietary product of Patton Consultants, Inc., are used here as illustrations.

Using the data entered on the equipment screen (see Figure 14-3), the computer program can quickly determine the next date or meter reading at which PM will be required. To do this, the logic algorithm divides the use per day into the PM interval. For example, a truck driven 65 miles per day divided into an interval of 7,500 miles produces an interval of 115 days. The program next considers the conversion between work days and calendar days. This could be seven-fifths if Monday through Friday is the normal work day and the truck sits idle Saturday and Sunday. The interval would be 168 calendar days when the six holidays are included. The computer then converts the Gregorian calculation into Julian calendar dates so that, for example, December 15th is the 349th day of 1981.

Adding $349 + 168 = 517$; since 1981 has 365 days, the calculation subtracts 365, giving the date next due as the 152nd day of 1982, which is translated to June 1, 1982.

Inspections and PMs based on the number of shifts can be easily adjusted by using a PM interval of shifts, and then the number of shifts per day can be one, two, or three.

It is important that the inspection or PM data be easily changeable. As much as possible should be accomplished automatically by the computer program. In COMMS, for example, meter readings on equipment may be updated either through the data entry at completion of each work order on that equipment or by a separate entry such as a driver's trip report. That meter information can be divided by the number of days to update the use per day continually, which then updates the next due date. When an inspection or PM is done and the work order closed, these data automatically revise the date last done, which again in turn revises the date next due.

An ability should be included in the computer program, probably through function keys on a "smart" CRT, to display the CRT screen for change data and rapidly jump to the desired field, make the change, and then close the program. Transaction logs of changes to PM intervals should be maintained for analysis that may indicate opportunity for improvement. The challenge here is that with an interactive computer system, data such as PM intervals may be easily changed, but the previous data and any trends will often not be maintained or remembered unless the human analyst recalls past changes. Computer power provides opportunity for condition monitoring maintenance if the needs are properly programmed.

ON-CONDITION MAINTENANCE

To date, few instruments are directly connected to the maintenance computer. Microprocessors and sensors allow vibration readings and other nondestructive test (NDT) data to be recorded and analysed. Presently, these readings primarily activate alarm enunciators or recorders that are individually analysed. There are, of course, process-control systems in use today that have the capability of signaling the need for more careful inspection and preventive maintenance. These devices currently are cost effective only for high-value equipment such as turbines and compressors. The added complexity of tying such sophistication into maintenance computer systems often causes more problems than it is worth. Progress is being made, however, and by the time this book reaches its readers, such devices may already be cost effective. Trend analysis for condition monitoring may, of course, be assisted by computer records.

COMPUTERIZED PM SCHEDULING

The maintenance planner will normally sit down about mid-week and schedule all inspections and PMs for the following week. The screen shown in Figure 15-1 makes this easy. The planner simply enters the designation of the organization that he wishes to plan for, and the ending date of the time period. That date would normally be the Saturday of the following week. The computer displays the PMs that are due line by line, ranked with the highest priority (lowest number) first. The planner has the option of saying yes or no to doing that PM simply by keying a "Y" (New Line is Y by default) or "N" if the work order should not be created at this time. A work order will normally be created for every inspection, calibration, or PM that is to be due. The program can automatically reserve a work order number for each selected PM and will do a programmed switch to the work order screen when all jobs have been decided. The equipment and description data can automatically be inserted in the work order by the PM scheduling program with the information pulled from the equipment records, as discussed earlier.

RESOURCE COORDINATION

The maintenance planner should take the list of inspection and PM work orders to the production manager and determine when equipment can be made available for the desired maintenance. Given those constraints, the work orders can now be dispatched for specific dates, times, and even personnel. It is often useful to schedule PM for the first job of the day since that gets the maintenance mechanic off to a good start, usually on a cold machine, without having to disrupt production in the middle of a shift.

When the work order is being dispatched, using the

```
                        1 J    I N S P E C T I O N / P M       D U E

ENTER ORGN (O) OR EMPL (E): E     ENTER #: 1112             NAME:   BIANCHI

                          ENTER 'AS OF' DATE:     6/30/82
                                         SYSTEM
     PRTY     PROD    EQPT ID              ACCT #             NAME/LOC    I/PM   WO#
      4       DISC    2434                 C128         PLANT 2,MIS        A     3319
     12       PRTR    2232                 C123         PLANT 3,RM4        B     3334
     12       PRTR    2298                 C128         PLANT 3,STK RM     B     3335
     16       CMPR    2101                 C130         PLANT 2,MIS




SORTING SELECTED I/PM DUE ------>      14
DO YOU WANT TO CREATE A WORK ORDER (Y) / (N) ?:   Y
```

Figure 15-1
Screen image for creating Inspection/PM work orders.

screen illustrated in Figure 15-2, the computer can automatically check inventory and make sure that any necessary parts are available, either in the employee's stock or at the stockroom that supports him. If parts are not available, a message should so indicate and also print out a list for the stockroom of what parts are needed and should be ordered. This printout is shown in Figure 15-3.

The planner should have the ability to override the computer program edits and to decide to do the PM even though the parts are not available. If the parts are available, they should be printed on the work order copy that will be given to the employee and also on either a copy of that same work order or a separate pick list that is printed for the stockroom. If the stockroom has its own printer, then the pick list can be printed there directly. At the same time, the computer should transfer those PM parts from the "on hand" inventory to "on reserve," linked to that specific work order. When the parts are picked up at the

```
                1 D    S C H E D U L E / D I S P A T C H    W O

ENTER WO#:  2488              DESC:   LOADER 21 PM A + ECO 3261
                              ASSIGNED ORGN/EMPL#:   M-1234   MARSHAL
        TASK:   1-3           DESC:   ECO 3261 MOTOR


            #     PLAN    ASSIGNED   ORGN/EMPL      START   -    TARGET   -   COMPLETE
CRAFT     PERS   HR-MIN     ID#        NAME       DATE     TIME    DATE        TIME

MECH       2      6-0      4044      FELLICA      3/ 8/82  0700  3/ 8/82      1400
                          2771      FELDMANN     3/ 8/82  0700  3/ 8/82      1400
WELD       1      2-30     3520      BEALE        3/ 8/82  0800  3/ 8/82      1045
ELEC       1      3-30     5213      MORGAN       3/ 8/82  1030  3/ 8/82      1500


       DISPATCH                  PRINT WORK ORDER COPIES: 4
DATE       TIME                  PRINT ACKNOWLEDGEMENT COPIES: 1
3/ 5/82   1028                   PRINT PROCEDURE CHECK LIST COPIES: 1
```

Figure 15-2
Schedule and dispatch work orders.

stockroom, a computer program should transfer the parts from "on reserve" to that specific work order. Then finally, when the work is done, the parts can be transferred to the specific equipment on which they are to be installed. Necessary cost accounting can be accomplished at the same time. Unfortunately, all work does not get done on time, so a CRT screen or paper report similar to the Inspection/PM Overdue report shown in Figure 15-4 is required.

Capabilities should also be designed into the computer programs to indicate any other active work orders that should be done on equipment at the same time. Modifications, for example, can be held until other work is going to be done and the effect can be accomplished most effi-

```
            PICK-UP LIST OF PARTS NEEDED FOR WORK ORDER      PAGE:  1

ORDER  #:  2488        PAD #:    123333     TYPE:  PM     PRI:  12

ASSIGNED EMPLOYEE:                     EMPLOYEE  STOCKROOM:
   ID #:   4044                          ID #:  S123
   NAME:  FELLUCA                        NAME:  STOCKROOM
          MICHAEL
                                        ADDR:  BLDG 2, ROOM 115
                                               JIM PACKRAT, STOCKKE
                                PHONE  #:  (716) 334-2555


REQUIRED PARTS:   ( ANY PARTS NEEDED FROM THE STOCKROOM HAVE BEEN PUT
                  ON RESERVE FOR THIS PARTICULAR WORK ORDER.  THIS
                  SHEET MUST BE PRESENTED WHEN PICKING UP ALL OR SOME
                  OF THESE PARTS AT THE STOCKROOM. )

                                         QTY           QTY
                                         REQ.   QTY    AT
                                         FOR    AT     EMPL .    QTY
TASK        PN            DESCRIPTION     WO     EMPL   STKRM   SHORT

  1- 3   305021    GASKET    , NEOPHRENE,OIL RES    2      2
          610501    MOTOR     ,24V,7HP,BASE 18X18    1             1




    _____              _____
        SUPERVISOR                       STOCKKEEPER
```

Figure 15-3
Pick-up list of parts.

ciently. A variation on the same theme is to ensure that emergency work orders will check to see if any preventive maintenance work orders might be done at the same time. Accomplishing all work at one period of downtime is usually more effective than doing smaller tasks on several occasions.

Preventive maintenance procedures may be stored in the computer and printed at the same time as when the work order is dispatched. Most computers have standard software for word processing or at least a line edit capability that can be used to enter and revise procedures. While paperwork from a computer system should be kept to a minimum, a printed procedure checklist that can be signed by the inspector should help assure responsible accomplishment of the tasks. Single-point control over procedures is a big help, especially on critical equipment. The risk of pulling an obsolete procedure from someone's file drawer is greatly reduced. If all items on a procedure can-

```
   1 K   I N S P E C T I O N S / P M   O V E R D U E

ENTER (O)RGN or (E)MPL: O          CURRENT DATE:   3/16/82
      ENTER #: 123                 NAME: GENERAL MAINT

PRTY     EQPT ID  I/PM    WO#  TGT CMPL     DUE     EMPL #      NAME
  4     2411        A     3199  2/28/82   2/28/82 9321      PHILLIPS
 10     5137        C     4114  3/ 6/82   3/ 6/82 4044      FELLUCA
 16     5520        A     4111  3/15/82   3/15 82 2271      FELDMANN




SORTING TOTAL SELECTED WORK ORDERS ----->                    3
```

Figure 15-4
An Inspection/PM overdue report.

not be accomplished at one shift, the document can be passed to the next shift supervisor or held for completion until the next day. When completed, the work order would be closed out and that related information entered automatically onto history records for later analysis. Safety inspections and legally required checks can be maintained in computer records for most organizations without any need to retain paper copies. If those paper records must be maintained for some legal reason, then they should be microfilmed rather than kept in bulky paper form.

Computer power greatly enhances the ability to plan, schedule, and control PMs accurately.

16

Motivation

Motivation to do PM well is a critical issue. A little extra effort in the beginning to establish a PM program will pay large dividends, but finding those additional resources when there are so many "fires" to put out is a challenge. Like with most things that we do, if we want to do it, we can. Hersberg's two levels of motivation, as outlined in Figure 16-1, help us understand the factors that cause people to want to do some things and not be so strongly stimulated to do others. Paying extra money, for example, is not nearly as motivating as are demonstrated results that show equipment running better because of the PM and a good "pat on the back" from management for a job well done.

A results orientation is helpful. This is because, as shown in Figure 16-2, an unfilled need is the best motivator. That need, in reference to PM, is uptime on equipment, desire to avoid breakdowns, and opportunity to achieve improvement. The converse is failures and downtime with resulting low production and angry customer users.

PRODUCTION/MAINTENANCE COOPERATION

Some organizations, such as General Motors' Fisher Body Plant, have established a position of Production/Maintenance Coordinator. This person's function is to ensure that equipment is made available for inspections and preventive maintenance at the best possible time for both organizations. He is a salesman for maintenance. This is an excellent developmental position for a foreman or supervisor. One year in that position will probably be enough for most people both to learn the job well and to become eager to move on to duties with less conflict.

Other organizations make production responsible for initiating a percentage of work orders. At Frito-Lay plants, for example, the production goal is 20 percent. This target

Motivational factors (Job centered)	Achievement Recognition Advancement Growth potential Responsibility Work itself
Maintenance/ Hygiene factors (Peripheral)	Salary Job security Status Relations with supervisor Relations with peers Relations with subordinates Work conditions Company policy Technical supervision Personal life

Figure 16-1
The two-factor theory of motivation.

stimulates both equipment operators and supervisors to be alert for any machine conditions that should be improved. It tends to catch problems before they become severe, rather than allowing them to break down. The results appear to be better uptime than in plants where a similar situation does not occur.

EFFECTIVENESS

Productivity is made up of both time and rate of work. Many people confuse motion with action. Utilization, which is usually measured as percentage of productive time over total time, indicates that a person is engaged in a productive activity. Drinking coffee, reading a newspaper, and attending meetings are generally classed as non-productive. Hands-on maintenance time is classed productive. What appears to be useful work, however, may be repetitious, ineffective, or even a redoing of earlier mistakes.

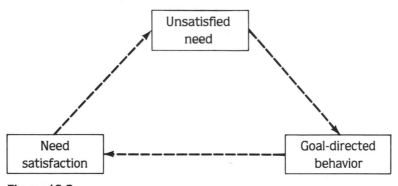

Figure 16-2
The process of motivation.

A technical representative of a major reprographic company was observed doing PM cleaning on a large duplicator. He spread out a paper "dropcloth" and opened the machine doors. The flat area on the bottom of the machine was obviously dirty from black toner powder, so the technical representative vacuumed it clean. Then he retracted the developer housing. That movement dropped more toner, so he vacuumed it. He removed the drum and vacuumed again. He removed the developer housing and vacuumed for the fifth time! On investigation, it was found that training had been conducted on clean equipment. No one had shown this representative the "one best way" to do the common cleaning tasks. This lack of training and on-the-job follow-up counseling is too common! To be effective, we must make the best possible use of available time.

There are few motivational secrets to PM, but there are guidelines that help:

1. Establish inspection and PM as recognized, important parts of the maintenance program.

2. Assign competent, responsible people.

3. Follow up to assure quality and to show everyone that management does care.

4. Publicize reduced costs with improved uptime and revenues that are the result of effective PM.

Implementing A New PM Program

Let's assume that you are starting from a base where you have virtually no preventive maintenance and most time is spent on emergency repairs. Let us imagine the process as like climbing a flight of stairs. You have made the first decision: to establish a PM program (climb the stairs), as shown on Figure 17-1.

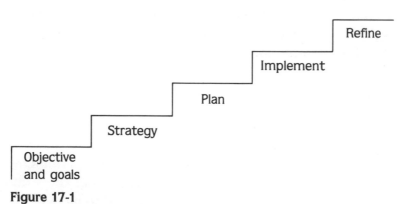

Figure 17-1
The climb to a new PM program.

OBJECTIVES AND GOALS

The next step is to detail objectives and goals. They should be (1) written; (2) understandable; (3) measurable; (4) challenging; and (5) achievable. The objective statement might include mention that the objective of the preventive maintenance program is to improve the organization's profits (effectiveness for nonprofit organizations) by increased equipment uptime, better performance of operating equipment, and reduced total maintenance costs. Typical goals could include:

1. Increase average availability of all production equipment from the present 92 percent, to 95 percent within six months, and 97 percent within twelve months.

2. Improve operating effectiveness, as measured by acceptable quantity units (pounds, gallons, activations, etc.) per hour; from the present period average of 1,435,000, up to 1,700,000 within six months and 1,850,000 within twelve months.

3. By improved preventive maintenance that should lead to reduced emergency maintenance and less production downtime, reduce the total costs of maintenance and lost production revenues from the present $450,000 per period, to $395,000 within six months and to $375,000 within one year, measured in constant dollars that exclude the effect of inflation.

PLANS

Strategy, as the next step, is generally to describe how the goals and objectives may be achieved. Once the strategy is determined, plans should be detailed to include the specifics of who, what, when, where, and how. As the plans are implemented, results should be evaluated, and revisions should be made as necessary to improve the PM program. Developing and implementing a good PM program will take a minimum of several months and could even be a year or more before all equipment is entered and the intervals and procedures are properly tuned.

If you start at the zero point, additional resources must be invested to get the program started. As PM activities begin to result in increased production and reduced emergency costs, the payback will increase to the point where benefits become greater than cost. Implementing a new PM program requires strong management support and direction to assure that the investment is being made in the right places, and that the payoff is achieved.

CHAPTER 18

Special Concerns

PARTS AVAILABILITY

Parts to be used for preventive maintenance can generally be identified and procured in advance. This ability to plan for investment of dollars for parts can save on inventory costs since it is not necessary to have parts continually sitting on the shelf waiting for a failure. Instead, they can be obtained just in time to do the job.

PM procedures should list the parts and consumable materials required. The scheduler should assure availability of those materials before the job is scheduled. Manually, this is done by checking inventory when the PM work order is created. The order should be held in a "waiting for resources" status until the parts, tools, procedures, and personnel are available. Parts will usually be the missing link in those logistics requirements. The parts required should be written on a pick list or copy of the work order given to the stock keeper. He or she should pull those parts and consolidate them into a specified pick-up area. It is helpful if the stock keeper writes that bin number on the work order copy or pick list and returns it to the scheduler so that the scheduler knows a person can be assigned to the job and production can be contacted to make the equipment available, knowing that all other resources are ready. It may help to send two copies of the work order or pick list to the stock keeper so that one of them can be returned with the part confirmation and location. Then, when the PMer is given the work order assignment, he sees on the work order exactly where to go to find the parts ready for immediate use.

It can be helpful, where specific parts are often needed for PM, to package them together in a kit. This standard selection of parts is much easier to pick, ship, and use, compared to gathering the individual item. Plugs, points, and a condensor are an example of an automobile tune-up kit, while air filters, drive belts, and disposable oilers are common with computer service representatives. Kits also make it easier to record the parts used for maintenance with less effort than the individual recording of piece parts. Any parts that are not used, either from kits or from individual draws, should be returned to the stockroom.

With a computer support system, parts availability can be automatically checked when the work order is dispatched. If the parts are not in the stockroom, the computer will indicate in a few seconds by message on the CRT screen that "All parts are not available; check the pick list." The printed pick list will show what parts are not on hand and what their status is, including availability with other personnel and quantities on order, at the receiving dock, or at quality-control receiving inspection. The scheduler can then decide whether the parts could be obtained quickly from another source to schedule the job now, or perhaps to place the parts on order and hold the work request until the parts arrive. The parts should be identified with a work order so that receiving personnel know to expedite their inspection and shipment to the stockroom, or perhaps can be shipped directly to the requiring location.

A similar capability should be established for parts that are required to do major overhauls and unique planned jobs. Working with the equipment drawing and replaceable parts catalog, one should prepare a list of all parts that may possibly be required. Failure-rate data and predictive information from condition monitoring should be reviewed to indicate any parts with a high probability of need. Parts replaced on previous, similar work should also be reviewed, both for those that obviously must be replaced at every tear down, and for those that will definitely not be replaced because they were installed the last time.

Once the list of parts needs is established, internal inventory should be checked and any parts that are available should be staged to an area in preparation for the planned work. Special orders should be placed for the additional required parts, just as they are placed to fill any other need.

REPAIRABLE PARTS

Repairable parts should receive the same kind of advance planning. If it can be afforded as a tradeoff against reduced downtime, a good part should be available to install and the removed repairable parts should be rebuilt at a later date and then restocked to inventory. If a replacement part cannot be made available, then at least all tools, fixtures, materials, and skilled personnel should be standing by when the repairable part is removed.

The condition of repairable parts, as well as those that are throw-aways, should be evaluated as soon as convenient. The purpose is to measure the parameters that could lead to failure and determine how much longer the part could be expected to operate without failure. If examination shows that considerable life is left on the part, then the PM or rebuild interval should be extended in the future. Removed repairable parts should be tagged to indicate why they were removed. There is nothing more frustrating to a repair person then trying to find a defect that does not exist.

DETAILED PROCEDURES

This topic has been covered earlier but should be re-emphasized to ensure that the best balance is developed between details and general functions. The following are some general guidelines:

1. Common words in short sentences should be used, with a reading comprehension level no higher then seventh grade.

2. Illustrations should be used where possible, especially to point out critical measurements.

3. Commonly done tasks should be referred to by function, whereas those tasks that are done once a year or less frequently may be described in detail.

4. Daily and weekly checklists should be protected with a transparent cover and kept on equipment.

5. Inspections and maintenance done once a month or less often should be issued as specific work orders.

6. The PMer's signature should be required on every completed job.

7. Follow-up inspection should be done by management on at least a large sample of the jobs in order to assure quality.

8. Failure rates on equipment should be tracked to increase inspection and PM on items that are failing, and to decrease efforts where there is little payoff.

9. What was done and how much time it took should be recorded as guidance for future work.

QUALITY ASSURANCE

Quality of maintenance is a subject that requires more emphasis than it has received in the past. Like quality of any product, maintenance quality must be designed and built in. It cannot be inspected into the job.

The quality of inspection and PM starts with well-designed procedures, equipment, and a surrounding environment that is conducive to good maintenance and management emphasis. The procedures must then be followed properly, adequate time provided to the PMer to do the job well, and standards available with training to illustrate what is expected. There is one best way to do most inspections and preventive maintenance. That way should be procedurized and controlled to assure successful completion.

First-line supervision is critical. Foremen should spend the majority of their time managing their people at the work site and assuring that customers are satisfied. It is not possible to manage preventive maintenance from behind a desk. A foreman must get out and participate in the jobs as they are being done and inspect them on completion. This motivates his people to do both high-quantity and high-quality work. He will be on the site to apply corrective action as needed and to provide final job inspection and close out the work order.

AVOIDING CALLBACKS

"Callbacks" are generally defined as any repeat requirements for maintenance that may result from problems that should have been alleviated earlier or that were caused by earlier maintenance. Some organizations define a callback as any emergency maintenance on the same equipment within twenty-four hours for any reason. Other organizations narrow to the same problem but within time periods as long as thirty days. A measure should be chosen that suits the specific type of equipment. If your organization services for pay, you certainly should not charge additional for callback service since the problem should have been fixed the first time.

The fact remains that highly reliable equipment is often serviced by low-reliability people. Preventive maintenance often incurs exposure to potential damage. The same steps that improve quality assurance also reduce the incidence of callbacks:

1. Establish and follow detailed procedures.
2. Train and motivate persons on the importance of thorough PM.
3. If it works, don't fix it.
4. Conduct a complete operational test after maintenance is complete.

REPAIRS AT PM

Two philosophies exist on the best way to handle repairs that are detected during preventive maintenance. One approach is to fix everything as it is discovered. The other extreme is to repair nothing but rather mark it on the work order and assure that follow-up work orders are created. A policy that falls between the two is recommended: fix the minor things that can be most quickly done while the equipment is available, and identify other problems for separate work orders. A guideline limit of ten minutes has proved useful to separate tasks that should be done at the time, from those that should be scheduled separately. Naturally, any safety problem that is found should result in shutdown of the equipment and be repaired before the equipment is operated again. Restricting the amount of repair done on PM work orders helps to control the PM activities so they can stay on schedule. Table 18-1 outlines the criteria to be considered for repair with PM versus separate repair.

It can thus be seen that a small work force with multi-skilled persons servicing equipment that requires long travel, has delay time to get on the equipment, and exten-

Table 18-1
Criteria for PM Repair Method

Repair Separate from PM	*Repair with PM*
Enables more accurate scheduling of PM, at consistent times.	Best if: Equipment is difficult to get from production.
Allows use of inspection specialists with separate repair expert.	Extensive tear down is involved that would have to be repeated for separate repairs.
Allows parts, tools, and documents to be obtained as required, instead of carrying extensive inventory.	Extensive travel time is required to return to location.
	It is difficult for the person discovering the problem to describe it to another repair person.

sive preparation and access time should do repairs at the same time as PM. If, however, the work force is large enough to become specialized and supports large numbers of similar equipment that are located close together, then the inspection/PM function should be separated from repairs. As a generality, most manufacturing plants should do repairs separately from PM. Most field service will do both at the same time.

DATA GATHERING

Maintenance management needs data. Maintenance personnel do not like to report data. Given this disparity between supply and demand, everything possible should be done to minimize data requirements, make data easy to obtain, and enforce accurate reporting. The main information needed from inspection and PM is:

1. That the job was done
2. Equipment use meter reading
3. Part numbers of any parts replaced
4. Repair work requests to fix discovered problems
5. Time involved.

As the PM sophistication increases toward predictive maintenance, the test measurements should be recorded so that signature and trend analysis with control limits can be used to guide future maintenance actions.

SUMMARY

The following points bring together some of the main points in the preceding discussions:

1. Preventive maintenance is necessary for most durable hardware.

2. PM enables preaction, which is better than reaction.

3. It is necessary to plan ahead.

4. A good data-collection and information-analysis system must be established to guide efforts.

5. All possible maintenance should be done at a single access.

6. Safety must be regarded as paramount.

7. Vital components must be inspected.

8. Anything that is defective must be repaired.

9. If it works, don't fix it.

True or False Questions

These 250 questions serve several purposes:

1. Review of main points in text and emphasis on priority topics.
2. Sharpen thinking on subtle differences.
3. Assurance that student has learned information.
4. Reveal weak areas of knowledge so they can be improved.
5. Grade for academic courses.

The answers are on separate pages at the end.

There are ten questions for the Preface and each chapter. They are sequenced so that the Preface is numbers 0–9, Chapter 1 is 10–19, Chapter 2 is 20–29, and so on.

There are a total of 180 chapter questions and 70 additional questions that cover the entire subject matter.

PREFACE

0. T / F   Every electromechanical device will fail some
             time.

1. T / F   Preventive maintenance includes actions in-
             tended to avoid future failures.

2. T / F   Maintenance technology and management
             are, fortunately, not changing very much.

3. T / F   If preventive maintenance can keep equip-
             ment working as well as when it was pur-
             chased, the capital investment is preserved.

4. T / F   Doing maintenance on equipment may fix
             one problem but cause others.

5. T / F   Management oriented to short-term profits
             will be enthusiastic about starting a PM pro-
             gram.

6. T / F   The need for facts to guide PM dictates the
             need for a good manual or computerized
             data system.

7. T / F   Once established, a PM program should be-
             come a stabilized, rigid process.

8. T / F   Facts have little influence on human percep-
             tions.

9. T / F   Sophisticated technology can eliminate the
             need for management direction.

CHAPTER 1

10. T / F   The three major types of maintenance are improvement, preventive, and corrective.

11. T / F   Modifying equipment so that it can be serviced with the same tools and test instruments that are already in use is an example of improvement maintenance.

12. T / F   On-condition maintenance means to monitor statistical trends and perform actions on all equipment when a failure pattern is detected.

13. T / F   UM is an abbreviation for breakdowns that require emergency repair and are classed as unscheduled.

14. T / F   The best kind of maintenance eliminates or at least reduces the problems that cause the need for maintenance actions.

15. T / F   Humans generally rise to the challenge of doing maintenance well under difficult conditions.

16. T / F   The majority of present maintenance is done by reacting to fix things after they have broken.

17. T / F   The PM technical challenge is to detect potential problems before they cause a major failure.

18. T / F   Condition monitoring is most effective where failure is sudden.

19. T / F   Equipment should be disassembled and inspected internally as often as labor resources will permit.

CHAPTER 2

20.   T / F   It is not possible to estimate PM times accurately because of great variations that should be encouraged for different people doing the same PM tasks.

21.   T / F   The word "preactive" implies that PM can be planned and action taken before any failure.

22.   T / F   Workload peaks that exceed available repair capacity become an even bigger problem if PM is emphasized.

23.   T / F   Maintenance organizations emphasizing PM will use far fewer total labor hours than they would if most maintenance were corrective.

24.   T / F   Improved equipment uptime, availability, and productivity is a major benefit from PM.

25.   T / F   There are generally several equally good ways to do a PM task.

26.   T / F   Along with the advantages for doing preventive maintenance, there are potential disadvantages.

27.   T / F   An active PM program requires larger quantities of spare parts than are required to support breakdown maintenance alone.

28.   T / F   It is better to keep quiet about a PM program so as not to upset people, rather than to publicize the efforts.

29. T / F   Production cooperation has little influence
           on a PM program.

CHAPTER 3

30. T / F   Past equipment failures should be analyzed so that the problems and causes can be eliminated.

31. T / F   Codes for problems, causes, and actions facilitate reporting and data analysis.

32. T / F   A meter that records use on major equipment components aids in providing information to establish optimum use intervals between PM.

33. T / F   Wear on a motor vehicle's tires will be precisely proportional to the number of miles on the odometer.

34. T / F   The initials FMECA stand for Failure Modes, Effects and Criticality Analysis, which is a method of determining what problems may be caused, what results are expected, and how important they are.

35. T / F   Maintenance personnel should be made responsible for all equipment, cleaning, lubrication, and adjustments.

36. T / F   The skills necessary for doing good PM can be easily obtained through formal education programs and immediately applied on the job.

37. T / F   If a team of persons is used to service equipment, the speed and quality of repairs are generally poorer than when single individuals are assigned the responsibility.

38. T / F   Maintenance tasks are done best if they can be completed without interruption.

39. T / F   Even with checklists, some persons notice and do a better job on some items than they do on others.

CHAPTER 4

40.   T / F   A 30 percent ROI means that your investment will only be reduced by 30 percent.

41.   T / F   Requirement for a one-year payback period may eliminate many PM projects that could have high ROI.

42.   T / F   Nonuniform investments must be calculated as single investments that are then summed together.

43.   T / F   A payback of 1 percent per month for one year is greater than 12 percent a year compounded annually.

44.   T / F   Capital recovery factors include payment of both interest and the principal.

45.   T / F   If a service manager desires to calculate how much a payback of $20,000 received at the end of each of the next three years is worth today, the factor is found in the Tables for Future Value of an Annuity in Arrears.

46.   T / F   Financial expertise can help with calculations, but maintenance service management must be the source of accurate, valid data.

47.   T / F   The least cost for maintenance will be the highest possible sum of corrective maintenance costs, preventive maintenance costs, and value lost from downtown.

48. T / F   It is easy to determine the optimum balance of corrective and preventive maintenance even without work order cost data.

49. T / F   If four PM jobs took 2.5, 3.0, 2.0, and 2.5 hours, the mean preventive time was 3.0 hours.

CHAPTER 5

50. T / F   Body senses are efficient detectors of large performance deviations.

51. T / F   Functional tests that evaluate performance cannot adequately indicate the need for maintenance.

52. T / F   If a moving component is disassembled from its mate, such as a brake shoe from its brake lining it is very difficult to fit them back into the same configuration and therefore additional wear will be induced.

53. T / F   Standards for go/no go conditions will greatly improve determination of whether or not to replace components.

54. T / F   Hardware that measures conditions and transmits information to external indicators is called a perceiver.

55. T / F   A base line "signature" for moving mechanical equipment should be recorded under known good conditions for comparison to later recordings.

56. T / F   The presence of metallic elements in a lubricant can indicate what components are wearing and how soon they may need to be replaced.

57. T / F   A safety margin calculated around a failure point gives confidence that equipment can be stressed beyond that point with very little risk of failure.

58. T / F   It is critical that a distinction be made between maintenance feedback controls and operational controls.

59. T / F   Putting a filter in the air intake of a computer cabinet unnecessarily adds another component that can become clogged with dirt and fail.

CHAPTER 6

60. T / F Most equipment operators are very attentive to the need for maintenance and carefully adhere to recommended intervals.

61. T / F The characteristic bathtub reliability curve is typical of most components in equipment used today.

62. T / F Components that incur infant mortalities which can be weeded out by burn in, and then stabilized for the rest of their lives, are best maintained by fixed interval replacement.

63. T / F It is possible for components to have a consistent failure rate over use.

64. T / F Increasing failures over equipment life indicate that burn in was not adequate.

65. T / F A sudden change, such as lighting of the red "low oil" indicator on a car's dash panel attracts an operator's attention more reliably than do variable gages.

66. T / F Vehicle oil changes done precisely every 3,500 miles are better for equipment life than changing the oil when it wears out.

67. T / F The most common failure pattern for electric components is early failures and then a stable life.

68. T / F   Exposure to many different operators can cause more failures after equipment is in widespread use than occurred during early use.

69. T / F   PM for most equipment is best done by inspection and lubrication with interior access only when conditions indicate the need.

CHAPTER 7

70. T / F Condition monitoring does not require as sophisticated a data-collection system as does on-condition maintenance.

71. T / F Economic justification of condition monitoring maintenance requires consideration of all labor, materials, and production costs to choose the least-cost alternative.

72. T / F Low labor costs and high parts costs will generally motivate a decision toward mass replacement of components.

73. T / F If all lights in a factory ceiling are replaced at one time, the parts cost will be the minimum possible.

74. T / F When production demands permit, it is best to accomplish all pending maintenance activities during a single access to the equipment.

75. T / F If equipment is very reliable and failure conditions build up slowly, the best method of PM is to wait for breakdown or a convenient visit to the equipment; and then inspect all critical components and do any maintenance dictated by problems found during the inspection.

76. T / F Detection of a failure pattern should result in preventive measures to eliminate further failures.

77. T / F Problem/cause/action analysis of past failures is not a reliable guide for improvement maintenance.

78. T / F Decision rules may be established in advance to guide when conditions should result in maintenance action.

79. T / F It is critical that PM efforts be categorized properly as on-condition, condition monitor, or fixed interval.

CHAPTER 8

80. T / F   Doing PM at the exact time planned is not always possible because of constraints on equipment availability for maintenance.

81. T / F   The two main elements of fixed interval PM are procedure and discipline.

82. T / F   The first element of PM planning and control is to list all equipment and the intervals at which it must receive preventive attention.

83. T / F   Quality-control inspections are not necessarily for maintenance.

84. T / F   Spending money to clean, repair, and refurbish seasonal equipment for storage indicates poor management, since the financial expenditures should be delayed until the equipment is needed again.

85. T / F   A master schedule should detail tasks only by annual, quarterly, and monthly categories.

86. T / F   People will be eager to do PM tasks, so there is no need to assign responsibility to a specific person.

87. T / F   Some PM tasks are best done by establishing a routine such as "first thing every Monday."

88. T / F   To be helpful, use meters on equipment must relate the measure with wear on the assemblies to be maintained.

89. T / F   Equipment used in frequent stop-and-go conditions will need less maintenance service than similar equipment running continuously.

CHAPTER 9

90. T / F The rubbing of two materials against each other causes deterioration due to heat and wear.

91. T / F Tribology is the study of rubbing and friction.

92. T / F The resistance of materials to wear has changed little over recent years, and probably will not change much in the future.

93. T / F Air and nitrogen can be used as lubricants.

94. T / F A technically sound lubrication design will not be influenced by management attention.

95. T / F Equipment manufacturers are one of the best sources of lubrication guidance.

96. T / F Lubricant vendors are eager to assist potential customers in selecting the best lubricants to be used on specific equipment.

97. T / F While it is possible to underlubricate components, overlubricating is a more common error.

98. T / F Additives are mixed with base oil in order to provide lubricant adherence and strength under conditions of heat, pressure, and rapid movement.

99. T / F Assigning a specific maintenance supervisor as responsible for implementation and operation of a lubrication program is not necessary since a good program should run itself.

CHAPTER 10

100. T / F Assuring that instruments meet criteria, normally derived from the National Bureau of Standards, is called calibration.

101. T / F Primary standards are available in most industrial tool rooms and development laboratories.

102. T / F Instruments should be checked for accuracy based strictly on intervals of time.

103. T / F It is helpful to retain, information on what actual readings are at each calibration, in order to set the proper interval between inspections.

104. T / F The inspector should initial or sign the calibration log.

105. T / F Placing a calibration label on each affected equipment places responsibility for assuring compliance directly onto the central calibration organization.

106. T / F A computer is mandatory for maintaining calibration records.

107. T / F If a volt/ohm meter is dropped, it should be checked for accuracy and recalibrated if necessary even though it is not yet due for recalibration.

108. T / F   If every instrument brought in for calibration is within control limits, the interval between calibrations should be extended.

109. T / F   A good calibration system helps avoid performance deficiencies and potential liabilities.

CHAPTER 11

110. T / F   PM procedures should detail, who, when, where, what, how, and why.

111. T / F   Listing tools, parts, and reference documents on a procedure helps the worker assure that all necessary resources are available.

112. T / F   Even the most experienced worker should at least occasionally use the PM procedure checklist to assure that every item is covered.

113. T / F   Central control over PM procedures has the major advantage of uniformity and assurance that any changes will be rapidly available to any user.

114. T / F   Any equipment found in unsafe condition may be kept in use if repairs are already on the work schedule.

115. T / F   Experience with similar equipment is the best means of determining time estimates.

116. T / F   It is possible to plan and estimate time for only a small percentage of PM work.

117. T / F   Labor costs are estimated by calculating the hours required times the applicable skilled-labor rates.

118. T / F Work that requires more than one person tends to be more efficient than single-person work.

119. T / F Fortunately, maintenance tasks do not generally follow the axiom that "work expands to fit the time available."

CHAPTER 12

120. T / F At a nuclear power station, the balance between labor costs, parts costs, and production revenues should have the major emphasis placed on reducing parts costs.

121. T / F One of the best tools for planning shutdowns is the critical path method (CPM).

122. T / F Wall charts provide graphic planning and
communications but require far more effort
than they are worth.

123. T / F As computer scheduling routines are developed, all possible efforts should be made
to eliminate schedule wall boards.

124. T / F Maintenance should direct production on
when shutdowns must be scheduled.

125. T / F Staging required parts to specific areas in
the stockroom or where they will be used
makes more work for stock keepers and is
little help to craftspeople.

126. T / F Future planning can be improved by com-
paring actual and estimated times and costs,
and striving for improvement.

127. T / F Parts removed during shutdown maintenance should be examined to determine
how much use they can withstand before
failure.

128.   T / F   The use of specialists to plan sequence, materials, and labor will generally slow down the process and lengthen shutdown time.

129.   T / F   Problem/cause/action analysis of past failures can help identify items that should receive special attention during shutdown.

CHAPTER 13

130.   T / F   The objective of scheduling is to balance demand for PM with the supply of resources.

131.   T / F   Customer rank should have no influence on the priority of work.

132.   T / F   In rating systems, people generally think of number one as the first item to be addressed.

133.   T / F   FIFO is the best method of prioritization since every requestor has to wait an equal time.

134.   T / F   Callbacks to redo unsatisfactory previous work should receive attention only after other work gets done.

135.   T / F   Need urgency and equipment criticality are directly related to the effect on production and revenues.

136.   T / F   Rating of customer (user) rank and equipment criticality should be done unilaterally by maintenance management.

137.   T / F   Computer-calculated priorities should not be subject to overriding decisions by humans.

138.   T / F   Proper prioritization will help assure that PM is done on time so that equipment keeps operating and emergency work is reduced.

139. T / F   Maintenance planning and production planning can together work out the general schedule for PM. Final arrangements will often be established by the maintenance person who is to do the PM contacting the first-line production supervisor in charge of the equipment to establish specific details.

CHAPTER 14

140. T / F A record system that keeps track of all open work on equipment or at a user location will be helpful to ensure that all resources are ready when the equipment becomes available or the level of problems dictates immediate maintenance.

141. T / F Supervisors who frequently check the quality of preventive maintenance work after the PM person reports that it is complete will harm the morale of their people.

142. T / F Repair, PM, and rebuild records should not be kept for every specifically identifiable piece of eqiupment.

143. T / F Use meters on equipment help to do PM at the proper time even though equipment operation may vary.

144. T / F The amount of information entered and retained in the equipment records depends on benefit/cost analysis for the expected future usefulness of the information versus the people, time, and cost of entering and retaining it.

145. T / F As an organization grows, equipment work records can be transformed from a log book to individual record cards to a computer.

146. T / F Identification of every equipment by a unique number is rarely needed for PM planning or control.

147.   T / F   A single interval should be set for PM of all equipment components.

148.   T / F   Having the location of equipment listed on records will help in planning the most efficient route of travel.

149.   T / F   Records should be kept of when every ceiling fluorescent light tube is replaced.

CHAPTER 15

150. T / F   Humans can do numerical calculations far more rapidly and accurately than computers can.

151. T / F   The number of hours a machine operates on the average day is calculated by dividing the total hours run by the number of days operated.

152. T / F   Computerized trend analysis will aid condition monitoring.

153. T / F   The number of days between PM activities should be the same on a numerically controlled machine whether it operates one, two, or three shifts; or five, six or seven days a week.

154. T / F   Once PM intervals and operating rates are set, they should not be changed.

155. T / F   Modern interactive computer systems can change data instantly, so transaction logs of changes to PM intervals and results of condition monitoring inspections should be kept, which can lead to improved PM.

156. T / F   Procedure documents are best kept as paper forms in a file drawer, rather than electronically in the computer.

157. T / F   Linking process control instruments and remote sensors into the central maintenance management computer is relatively easy and reliable.

158. T / F   The scheduling of PM work is constrained by access to the equipment, parts, and personnel availability.

159. T / F   Electronic data in the computer satisfy most legal inspection record requirements.

CHAPTER 16

160. T / F   Personal satisfaction with seeing positive results from PM activity is a strong motivator.

161. T / F   The position of production/maintenance coordinator can be an excellent learning experience and a positive contributor in spite of inherent conflict.

162. T / F   Production personnel should not request work orders since only maintenance personnel know when equipment is having problems.

163. T / F   Utilization is a percentage of total time divided by productive time, so it will always be 100 percent or more.

164. T / F   The time required to weld a new tool rack onto an NC machine should be classed as nonproductive.

165. T / F   It is possible to look busy and be engaged in direct repair activity that is in fact not productive.

166. T / F   Management recognition that inspection and PM are important generally motivates personnel and results in a self-fulfilling prophecy so that equipment is better maintained and runs better as a result.

167. T / F   Paying a bonus for improved equipment uptime is the best possible motivator.

168.   T / F   There is "one best way" to do most maintenance tasks.

169.   T / F   On-the-job counseling is a sign that management considers the activity unimportant until it is "fouled up."

CHAPTER 17

170. T / F   The first step in establishing a new PM pro-
             gram is to write objectives and goals.

171. T / F   A manager should be careful that goals are
             not accurately measurable so that respon-
             sibility can be avoided.

172. T / F   Goals should be challenging so that people
             have to exert additional effort to reach
             them, but achievable so that the people are
             satisfied when they are reached.

173. T / F   Strategy describes how the goals and objec-
             tives may be achieved.

174. T / F   Establishing a good PM program can be
             done in a few weeks if management pro-
             motes it forcefully.

175. T / F   A well-planned PM program will rarely need
             revisions.

176. T / F   If an organization is starting with little for-
             mal PM, then initial costs to start the pro-
             gram will be higher than initial benefits.

177. T / F   Quantifying goals in constant-year econom-
             ics makes performance measurement more
             realistic since the effect of inflation is re-
             moved.

178. T / F   Improved profits and effectiveness should
             not be included in objectives of a new PM
             program.

179.   T / F   A new PM program implemented where none previously existed should improve equipment availability by several percentage points.

CHAPTER 18

180. T / F   Packaging a standard selection of parts in a kit guarantees that all parts will be used.

181. T / F   A category of work "waiting for resources" can be effective to make those work orders visible and assure that they are not scheduled until required parts are on hand.

182. T / F   Parts that were drawn from stock but not needed on equipment should be scraped.

183. T / F   Even though a machine component can be repaired, it is often better to have a good replacement available for immediate replacement, and repair the removed component later.

184. T / F   Examination of removed parts that indicates the part was not yet worn near failure should result in the replacement interval being shortened.

185. T / F   Requiring the person doing the job to sign off when it is done helps assure quality PM.

186. T / F   Procedures should cover common tasks in detail, and tasks done infrequently need only be covered in generalities.

187. T / F   Call backs provide a good measure of maintenance quality control.

188. T / F   After maintenance is complete, equipment should not be operationally tested for fear of causing further problems.

189. T / F   Recording of test measurements aids in trend analysis for guiding future maintenance actions.

GENERAL QUESTIONS

1. T / F   A PM program usually pays off the initial investment almost immediately.

2. T / F   Humans are favorably stimulated by designing some difficult-to-maintain components into every equipment.

3. T / F   The need for repairs should be completely eliminated by PM.

4. T / F   An obvious failure is the easiest kind to diagnose.

5. T / F   Automatic detection devices can set threshold standards at which maintenance must be performed.

6. T / F   Safety is a major consideration that drives the need for preventive maintenance.

7. T / F   Scheduled, fixed interval maintenance is generally less effective and not as economical as on-condition and condition monitoring maintenance.

8. T / F   Replacing a light bulb when it burns out is an example of condition monitor maintenance.

9. T / F   Any advantages and disadvantages will be the same regardless of the type of PM used.

10. T / F   Persons are usually slow in learning PM activities because of the great variation in requirements from time to time.

11. T / F   One function of inspections is to assure that equipment does not become unsafe.

12. T / F   Bottom-line profits should be an organization's objective.

13. T / F   Investment costs may be increased if the benefits are greater than the costs, and the results are improved productivity and profits.

14. T / F   Callbacks and problems caused by inept maintenance are extremely rare.

15. T / F   "Infant mortality" means that many parts will fail during early use.

16. T / F   The longest possible part life is obtained by using that part until failure.

17. T / F   An investment in PM is being made in future dollars and will be repaid in current dollars.

18. T / F   It is possible to do too little PM, but not to do too much.

19. T / F   Standardization of maintenance codes should be forced so that they are the same for all organizations and equipments.

20. T / F   Forcing grease through a dirty fitting will lubricate the components just as well as if the fitting were cleaned first.

21. T / F   If a person normally operates the same equipment and can be properly trained, then he should do his own lubrication.

22. T / F   PM should be relegated to the status of un-important, undesirable work.

23. T / F   The human senses of an experienced person are one of the best detection systems available today.

24. T / F   Rotating the PM responsibilities allows a person who does PM one week to benefit from improved equipment performance during following weeks when he or she is on corrective maintenance duty.

25. T / F   The best maintainability eliminates the need for maintenance.

26. T / F   Forecast accuracy is generally much better in the near future than it is over the long term.

27. T / F   Since payback in the future will be greater than in the present, the Future Value Table factors are all smaller than one.

28. T / F   Supporting equipment with 100 percent of all maintenance done preventive is a virtual impossibility.

29. T / F The cost of maintenance is directly related to the amount of labor and parts used.

30. T / F A satisfactory low-cost combination of PM/CM/downtime will be more realistically achievable than the optimum balance point.

31. T / F Doing more PM than the minimum cost (optimum) point will increase total costs, but may have offsetting benefits.

32. T / F A PM program should be stopped if improvements are not apparent within two months.

33. T / F Starting a PM program will cost additional money since the corrective maintenance must be initially continued as well as the new PM.

34. T / F The difference between MTBF and MTBM, is PM.

35. T / F Operational availability (Ao) includes getting all items for and performing PM as well as CM.

36. T / F Time elapsing between occurrences of an event has little impact on human sensitivity to any difference.

37. T / F The analytical process known as "SOAP" measures spectral performance of both fluorescent and incandescent lights.

38. T / F It is helpful to induce failures in equipment under test conditions to know where the failure point really is.

39. T / F Replacing components before the end of their useful life will usually result in higher labor and parts costs.

40. T / F So-called "idiot lights" on automobiles replaced analog instruments solely on the basis of lower initial costs.

41. T / F The main function of a lubricant is to keep moving materials separate from each other.

42. T / F Equipment that is primarily electronic, such as computer peripherals, fortunately do not need lubrication.

43. T / F Quality rejects will be totally eliminated through a PM program.

44. T / F A maintenance policy of creating separate work requests for deficiencies found during PM is most effective for situations where field engineers have to travel long distances to get to the equipment.

45. T / F The PM inspector may often be a general practitioner, and repairs can best be made by a specialist.

46. T / F It is very difficult for an unskilled observer to detect whether or not a maintenance person is performing properly.

47. T / F Estimates should be prepared based on exact hours and minutes that should be required under perfect conditions, and then adjusted for expected and actual delays.

48. T / F Minutes are easier for maintenance personnel to understand and report than are decimal hundredths or even tenths of hours.

49. T / F Materials costs for consumables such as grease and waste rags should be lumped in overhead costs, but transmission oil can usually be a direct cost on specific equipment.

50. T / F It is possible to treat PM inspection and lubrication as an incentive operation based on both the time required to do the tasks and equipment uptime results.

51. T / F It is not generally possible to plan PM a day ahead.

52. T / F FIFO (First in First out) scheduling means that all work is done in a time sequence with the oldest work done next.

53. T / F "Criticality 1" equipments and "Rank 1" customers should be limited to 10 percent of the respective totals, since they will probably account for the majority of activity.

54. T / F Machine operators are very capable of detecting unusual equipment conditions.

55. T / F   PM must often be done on holidays and weekends when production personnel are on vacation.

56. T / F   PM cycle time means the number of hours equipment will be required from production.

57. T / F   If equipment needs monthly PM, doing one on May 30 and the next one on June 3 meets the practical requirement with minimum effort.

58. T / F   Repair calls should never be converted or expanded to add preventive maintenance.

59. T / F   It is important to program the computer so that maintenance planners do not have an option of deviating from the computer's calculations.

60. T / F   If parts normally replaced at PM are not available, the planner should be alerted, but one should be able to do the rest of the work anyway and report the exceptions.

61. T / F   A printed PM procedure checklist facilitates the transition from personnel on one shift to another and still assures that all work gets done.

62. T / F   There are many equally good ways to do most inspections and preventive maintenance.

63.   T / F   Knowledge of what parts are available, whether they be on hand in the stockroom, with other personnel, at the receiving dock, or at quality-control inspection, helps maintenance planning and scheduling.

64.   T / F   Staging parts needed for major maintenance activities to a separate area wastes effort and results in those parts being unavailable for emergencies.

65.   T / F   A repair tag indicating why a machine component was removed is nice for record keeping but has little value for repair personnel.

66.   T / F   Quality assurance is a production function that has little relevance to maintenance.

67.   T / F   Good planning will obtain parts for PM just in time to do the work.

68.   T / F   The most effective maintenance manager will operate mostly at his/her desk and computer terminal.

69.   T / F   The skills and abilities involved in inspection are the same as for repair.

70.   T / F   Maintenance personnel are usually very capable and eager to provide data needed by management.

Answers

CHAPTER 6

60.	F
61.	F
62.	F
63. T	
64.	F
65. T	
66.	F
67. T	
68. T	
69. T	

CHAPTER 9

90. T	
91. T	
92.	F
93. T	
94.	F
95. T	
96. T	
97.	F
98. T	
99.	F

CHAPTER 12

120.	F
121. T	
122.	F
123.	F
124. T	
125.	F
126. T	
127. T	
128.	F
129. T	

CHAPTER 7

70.	F
71. T	
72.	F
73.	F
74. T	
75. T	
76. T	
77.	F
78. T	
79.	F

CHAPTER 10

100. T	
101.	F
102.	F
103. T	
104. T	
105.	F
106.	F
107. T	
108. T	
109. T	

CHAPTER 13

130. T	
131.	F
132. T	
133.	F
134.	F
135. T	
136.	F
137.	F
138. T	
139. T	

CHAPTER 8

80. T	
81. T	
82. T	
83.	F
84.	F
85.	F
86.	F
87. T	
88. T	
89.	F

CHAPTER 11

110. T	
111. T	
112. T	
113. T	
114.	F
115. T	
116.	F
117. T	
118.	F
119.	F

CHAPTER 14

140. T	
141.	F
142.	F
143. T	
144. T	
145. T	
146.	F
147.	F
148. T	
149.	F

CHAPTER 15

150.   F
151.   T
152.   T
153.   F
154.   F
155.   T
156.   F
157.   F
158.   T
159.   T

CHAPTER 16

160.   T
161.   T
162.   F
163.   F
164.   F
165.   T
166.   T
167.   F
168.   T
169.   F

CHAPTER 17

170.   T
171.   F
172.   T
173.   T
174.   F
175.   F
176.   T
177.   T
178.   F
179.   T

CHAPTER 18

180.   F
181.   T
182.   F
183.   T

184.   F
185.   T
186.   F

187.   T
188.   F
189.   T

GENERAL QUESTIONS

1.		F	25.	T		49.	T	
2.		F	26.	T		50.	T	
3.		F	27.		F	51.		F
4.	T		28.	T		52.	T	
5.		F	29.	T		53.	T	
6.	T		30.	T		54.	T	
7.	T		31.	T		55.	T	
8.		F	32.		F	56.	T	
9.		F	33.	T		57.		F
10.	T		34.	T		58.		F
11.	T		35.	T		59.		F
12.	T		36.		F	60.	T	
13.	T		37.		F	61.	T	
14.		F	38.	T		62.		F
15.	T		39.	T		63.	T	
16.	T		40.		F	64.		F
17.		F	41.	T		65.		F
18.		F	42.		F	66.		F
19.		F	43.		F	67.	T	
20.		F	44.		F	68.		F
21.	T		45.	T		69.		F
22.		F	46.	T		70.		F
23.	T		47.	T				
24.	T		48.	T				

Selected Readings

ARTICLES AND PERIODICALS

The following are published since 1980. Earlier books, articles and periodicals are listed in the author's text, *Maintainability and Maintenance Management*, ISA, 1980.

Agresti, William W., "Managing Program Maintenance," *Journal of Systems Management*, Volume V33N2, February 1982, pp. 34–37.

Anonymous, "Right On The Money With Growing Service Bays," *Discount Merchandiser*, Volume V21N9, September 1981, pp. 84–86.

Anonymous, "Why We Computerized Our Truck Maintenance," *Modern Materials Handling*, Volume V37N12, August 6, 1982, pp. 48–51.

Anonymous, "Maintenance: Making a Clean Sweep Cheap," *Chain Store Age Executive*, July 1982, pp. 57–61.

Anonymous, "Test Equipment Keeps Pace With User Requirements," *Data Communications*, Volume V11N6, June 1982, pp. 62–64.

Anonymous, "How The Royal Manages and Controls Its Computer Network," *Canadian Datasystems* (Canada), Volume V13N1, November 1982, pp. 50–52.

Anonymous, "Contract Maintenance Program Boosts Uptime," *Production*, Volume V88N3, September 1981, pp. 79.

Anonymous, "Total Fleet Upgrading Yields a Bonus," *Traffic Management*, Volume V2N7, July 1981, pp. 38–41.

Anonymous, "Machine Maintenance: Easy Does It," *Chain Store Age Executive*, Volume V57N7, July 1981, pp. 108.

Blumenthal, Marcia, "Vendors Seek New Ways Of Maintaining Systems," *Computerworld*, Volume V15N20, May 18, 1981, pp. 1, 10.

Burger, Eugene J., "A Total Maintenance System," *Journal of Property Management*, Volume V46N6, Nov./Dec. 1981, pp. 325–331.

Feichtinger, Gustav, "The NASH Solution of a Maintenance-Production Differential Game," *European Journal of Operational Research* (Netherlands), Volume V10N2, June 1982, pp. 165–172.

Fine, David, "Maintenance Management And The Financial Executive," *Cost & Management* (Canada), Volume V56N1, Jan. Feb. 1982, pp. 36–41.

Gehring, H., Pachow, U., Zimmermann, H. J., Rokohl, H. J., "Reducing Radiation Exposure in Nuclear Power Plants by Appropriate Maintenance Policies," *European Journal of Operational Research* (UK), Volume V8N1, September, 1981, pp. 31–39.

Kavanau, Louis, "Preventive Maintenance for Daisywheel Printers," *Office*, Volume V95N6, June 1982, pp. 36, 40, 44.

Kull, David, "Staying Up 100%," *Computer Decisions*, Volume V14N2, February 1982, pp. 176–198, 234.

Manning, R. P., "Subscriber Line Testing-Directions For The Future," *Telephone Engineer & Management*, Volume V85N22, November 15, 1981, pp. 94–98.

Nguyen, C. G., Murthy, D. N. P., "Optimal Preventive Maintenance Policies For Repairable Systems," *Operations Research*, Volume V29N6, Nov. Dec. 1981, pp. 1181–1194.

Ottinger, Lester V., "Robot System's Success Based On Maintenance," *Industrial Engineering*, Volume V14N6, June 1982, pp. 38–40, 42–43.

Rhodes, Wayne, L. Jr., "Maintenance: Finding A Cheaper Way," *Infosystems*, Volume V28N5 (Part 1), May 1981, pp. 86–92.

Schwartz, Barbara, "How To Get The Most For Your Maintenance Dollar," *Output*, Volume V2N5, July 1981, pp. 40–45, 67.

Whitmore, Graham, "Environmental Control—Cooling It The Right Way," *Data Management*, Volume V19N7, July 1981, pp. 14–15.

NATIONAL TECHNICAL INFORMATION SERVICE ARTICLES

The following articles are available from National Technical Information Service (NTIS), 5285 Port Royal Road, Springfield, Virginia 22161.

AD-A111 352/1, "Techniques Suitable For A Portable Wear Metal Analyzer."

Conf-801115-35, "Fleet Servicing Facilities For Testing And Maintaining Rail And Truck Radioactive Waste Transport Systems."

PB-82-120288, "Development Of An Equipment Management System."

PB-81-225666, "The Service-Evaluated Products List For Rapid Transit Car Subsystem Components."

PB-81-970422, "Computer Program For Preventive Maintenance."

EPRI-AP-1610, "Guide For The Assessment Of The Reliability Of Gasification-Combined-Cycle Power Plants."

HEDL-SA-2211, "Preventive Maintenance And Load Testing Of Fixed Position Cranes In Support Of Major Operations."

PB81-167496, "Analysis Of Typical Vehicle Repair Costs, Phase II."

KAERI-196/RR-65/79, "Development Of Maintenance And Trouble Shooting For Nuclear Instrumentation System."

PB-80-209729, "Bridge Deck Deterioration: A Review Of New York's Experience."

PB-80-206469, "Maintenance Management And Service Contracts For Housing Manager." Participant's Workbook.

PB-80-197767, "Preventive Maintenance Management Manual For Fossil Steam Electric Generating Plants."

COO-4184-8, "Developing Maintainability For Fusion Power Systems." Final Report.

AD-A085 053/7, "A Model For Determining Optimal Preventive Maintenance Intervals For Tanks."

UNI-SA-61, "Using Quality Assurance During Maintenance And Planned Outages."

ANL/EES-TM-71, "Overview Of Relcomp, The Reliability And Cost Model For Electrical Generation Planning."

AD-AO78 606/1, "Estimating The Time Required To Transition Aircraft Fleets To New Scheduling Maintenance Intervals."

PB80-109838, "Study Of Alternative Approaches To Motor Vehicles Maintenance."

AD-AO75 589/2, "Improving The Preventive Maintenance Checks And Services Programs For The M60A1 Main Battle Tank."

PB-297 007/7SL, "Shipboard Maintenance And Repair System—Steam-Turbine Plant Prototype."

NTIS/PS-79/0284/4SL, "Mill Maintenance: General Mill Maintenance—Fires, And Explosions." (Citations from the Institute of Paper Chemistry).

INDEX

191